The Irreducible Tensor Method for Molecular Symmetry Groups

J. S. Griffith

Dover Publications, Inc.
Mineola, New York

Bibliographical Note

This Dover edition, first published in 2006, is an unabridged republication of the work originally published in 1962 by Prentice-Hall, Inc., Englewood Cliffs, New Jersey.

International Standard Book Number: 0-486-45047-3

Manufactured in the United States of America
Dover Publications, Inc., 31 East 2nd Street, Mineola, N.Y. 11501

Preface

This book gives an account of a particular version of the irreducible tensor method for molecular symmetry groups. It uses as its basic mathematical technique the theory of the V, W, and X coefficients which I have defined for finite groups in analogy with Racah's $\overline{V}$, $\overline{W}$, and X coefficients and Wigner's $3j$, $6j$, and $9j$ symbols. It is my hope and belief that these coefficients provide a rather convenient formalism for many-electron problems, which is why I have prepared a unified presentation of the irreducible tensor method for molecular groups in terms of them.

The book was written during my stay in the Chemistry Department of the University of Pennsylvania and I was supported in part by that department and in part by the Material Science programme of the University. I am happy to acknowledge the friendly hospitality of the chairman of the Chemistry Department, Dr. C. C. Price, the department, and the University.

J. S. GRIFFITH
FELLOW OF KINGS COLLEGE
CAMBRIDGE

Contents

3

V Coefficients for Other Symmetry Groups

Page 25

4

W Coefficients

Page 32

5

Irreducible Products and Their Matrix Elements

Page 40

6

Two-Electron Formulae for the Octahedral Group

Page 49

7

Fractional Parentage

8

X Coefficients

9

Spin

10

Matrices of One-Electron Operators

CHAPTER 1

Introduction

Quantum mechanical calculations for systems having symmetry may usually be divided fairly completely into two parts. One part consists of deriving as much information as possible from the symmetry alone. The other is the evaluation of certain integrals, the estimation of parameters, or the solution of equations which have no symmetry or for which symmetry considerations can provide no information. In most problems these two parts are, superficially at least, inextricably tangled. The irreducible tensor method is designed to separate them and then to provide a well-developed and highly efficient way to make use of the symmetry. It has been extensively developed for systems with Hamiltonians having full spherical symmetry (for example, free atoms or nuclei), and has been adequately described in books by Edmonds (2), Fano and Racah (3), Wigner (4), and Jucys, Levinsonas, and Vanagas (5).

In this book an analogous treatment is given for systems having the lower symmetry characteristic of the internal modes of motion of molecules and of the states of ions in solids. This treatment is of general applicability to molecular problems, but here we restrict ourselves almost entirely to the basic theory and to tables of the most important quantities occurring in it for the commonest groups. There are but a few examples, selected purely for illustration and chosen almost entirely from the theory of transition-metal ions.

The reader is assumed to have some familiarity with quantum mechanics and group theory. However, the book is fairly self-contained, although it relies at times on results quoted from Fano and Racah (3) or Griffith (26) but not proved here. This reliance is inevitable in a short book written at the present level.

The least sophisticated kind of application of group theory is to the determination of selection rules. With tetrahedral symmetry, for example, one may say that electric dipole transitions are forbidden between a pair of states transforming, respectively, as components of the A_1 and T_1 representations. This prohibition exists because the electric dipole moment transforms as T_2, and that representation is not contained in the direct product of the representations A_1 and T_1. However, such an argument discards much information implicit in the symmetry classification of functions and operators. It would say that transitions are allowed between A_1 and T_2 but would miss the fact that light polarized with its electric vector parallel to the OZ axis could not induce a transition between an s orbital and a p_x or p_y orbital. Furthermore, it would be unable to tell one that in spontaneous emission the three transitions

$$s \longrightarrow p_x, \qquad s \longrightarrow p_y, \qquad s \longrightarrow p_z$$

all have equal probability. The concepts of V coefficients and reduced matrix elements give the necessary extension of the simple group-theoretic argument mentioned above. They are defined, their properties are investigated, and tables of values are given for the common point symmetry groups in Chapters 2 and 3 and Appendices C and D. The irreducible tensor method is built with them as a foundation, and the rest of the book is devoted to the erection of the superstructure, first for the finite symmetry groups alone and then by using those groups for spatial symmetry and the Racah-Wigner formulation for the spin operators and functions. The formulae are often superficially complicated, but their derivation is always straightforward and their application easy. It will then appear that, in a fairly literal sense, the irreducible tensor method is able to extract and manipulate at will all the information in any problem which is implicit in the symmetry of that problem.

CHAPTER 2

V Coefficients for the Octahedral Group

2.1 INTRODUCTION

In this chapter we define V coefficients and derive many of their general properties.* We use the octahedral group without inversion, O, as an illustrative example and reserve the discussion of other particular groups until the next chapter. The group O has five inequivalent irreducible representations; we write them A_1, A_2, E, T_1, and T_2. Its character table is shown in Table 2.1. We choose the coordinate axes, OX, OY, and OZ, to be fourfold axes of the octahedron.

* The reader may refer to Griffith (26), Chapter 6, for a more leisurely introduction to the preliminary group theory.

Table 2.1. The character table for the group O. The fourfold axes are taken to coincide with the cartesian axes. mC_n denotes a class of m rotations, each of $2\pi/n$, about some axis.

O	1	$8C_3$	$3C_2$	$6C_2'$	$6C_4$
A_1	1	1	1	1	1
A_2	1	1	1	−1	−1
E	2	−1	2	0	0
T_1	3	0	−1	−1	1
T_2	3	0	−1	1	−1

Now we discuss the notation used for sets of functions which transform according to irreducible representations. Choosing the representation T_1 as an example, we observe that it has degree 3; this means that a set of linearly independent functions which forms a basis for T_1 must have just three members. We say they span T_1. One such set is furnished by the three components x, y, and z of the vector $\mathbf{r}$. The operation of the elements g of the group O on the row vector $\mathbf{r}$ gives a matrix representation $M(g)$ of the group O according to the rule

$$g\mathbf{r} = \mathbf{r}M(g)$$

Written out in detail, this equation would be

$$g[x, y, z] = [gx, gy, gz] = [x, y, z]\begin{bmatrix} M_{11}(g) & M_{12}(g) & M_{13}(g) \\ M_{21}(g) & M_{22}(g) & M_{23}(g) \\ M_{31}(g) & M_{32}(g) & M_{33}(g) \end{bmatrix}$$

The character of the element g in this representation is

$$\chi(g) = M_{11}(g) + M_{22}(g) + M_{33}(g)$$

In other words, it is the trace of the matrix $M(g)$. The set of $\chi(g)$ for all the elements g is called the character of the representation.

We shall actually be interested only in basis functions which form an orthonormal set. Such a set might be derived from the vector $\mathbf{r}$ by multiplying it by a suitable scalar function $f(r)$ and would give rise to identically the same matrix representation. Any other orthonormal set of functions f_i forming a basis for T_1 gives rise to a matrix representation $M'(g)$ of O having the same character but not necessarily the same matrices. However, M' must be related to M by the relation

$$M' = U^{-1}MU$$

for some unitary matrix U. Such a U is the same for all g, and the two matrix representations are said to be equivalent. It is always possible to choose linear combinations of the f_i which give identically the same matrix representation as does $\mathbf{r}$. When such combinations are formed, we shall write them f_x, f_y, and f_z and call them the x, y, and z components, respectively. These components are not determined completely by the original set of f_i; only their ratios are (23). If, however, we require them to be real and normalized, the only possible ambiguity remaining is that of a sign change throughout.

In the case of T_1, the analogy between our components and the cartesian components of a vector in the sense of elementary vector analysis is extremely close. Similarly, a function h_a which transforms as the unit representation A_1 may be usefully related to a scalar. However, whereas a true scalar is left unchanged by any rotation of the coordinate system, h_a is only necessarily unchanged by the 24 rotations which belong to the octahedral group.

The other three irreducible representations do not correspond quite so closely to concepts in vector analysis, but the idea of components applies in just the same way. For E components are so defined that the same matrix representation is obtained as is given by the pair of functions $2z^2 - x^2 - y^2$ and $\sqrt{3}\,(x^2 - y^2)$. For T_2 we use the three functions yz, zx, xy. For A_2 there is only one function, so there is no choice to make.

The preceding definitions are summed up in Table 2.2 together with

Table 2.2. The nomenclature for the components of irreducible representations with an example of a particular basis for each representation.

Representation	*Component*	*Example*
A_1	ι	$x^2 + y^2 + z^2$
A_2	ι	xyz
E	θ	$2z^2 - x^2 - y^2$
	ϵ	$\sqrt{3}(x^2 - y^2)$
T_1	x	x
	y	y
	z	z
T_2	x	yz
	y	zx
	z	xy

the nomenclature we shall use for the components. It may seem to the reader that it is unnecessarily confusing to use the same set of symbols for components for T_1 and T_2, as well as the same symbol for A_1 and A_2. This is not so; the reason will appear in Section 2.7.

We now consider the terminology which will be used for a general set of functions transforming according to an irreducible representation. We use lower-case English letters for representations and roughly corresponding Greek letters for components. Then the αth component of a set of functions f transforming according to the irreducible representation a will be written f^a_α. In ket notation such a component would be written $|a\alpha\rangle$, or perhaps $|N'a\alpha\rangle$, transforming according to the same representation. We always write the degree of a representation a, i.e., the number of independent functions, as $\lambda(a)$. The way in which Greek letters are taken to correspond to English is shown in the sequence:

$$|a\alpha\rangle, \quad |b\beta\rangle, \quad |c\gamma\rangle, \quad |d\delta\rangle, \quad |e\epsilon\rangle, \quad |f\phi\rangle, \quad |g\eta\rangle, \quad |h\theta\rangle, \quad |i\iota\rangle,$$
$$|j\xi\rangle, \quad |k\kappa\rangle, \quad |l\lambda\rangle, \quad |m\mu\rangle, \quad |n\nu\rangle, \quad |p\pi\rangle, \quad |r\rho\rangle, \quad |s\sigma\rangle, \quad |t\tau\rangle$$

Finally, so long as we use the system of quantization shown in Table 2.2 we shall insist that all spatial functions, kets, and bras are real, just as the matrices of the representations are. So a bracket expression $\langle f^a_\alpha \mid g^b_\beta\rangle$ is equal to both

$$\int f^a_\alpha g^b_\beta d\tau \quad \text{and} \quad \int g^b_\beta f^a_\alpha d\tau$$

and so also to $\langle g^b_\beta \mid f^a_\alpha\rangle$. On the other hand, we know that an operator p can have two kinds of conjugate complex, one being the conjugate linear operator $\bar{p}$ and the other the Kramers conjugate p^* [Section 2.6 and Kramers (7)]. The operators $\mathbf{l}$ and $\mathbf{s}$ for orbital and spin angular momenta play a large part in our analysis, and they cannot be taken to be real in both these senses simultaneously.

2.2 COUPLING COEFFICIENTS

If we have two sets of functions f^a_α and g^b_β, then the set of products $f^a_\alpha g^b_\beta$ span the direct product ab of the two constituent representations. The representation ab may be reducible. Write $ab = \Sigma\, c$ where the c are irreducible. Then we know for O, if a and b are irreducible, that no representation can occur more than once among the c.

Certain linear combinations h^c_γ of the $f^a_\alpha g^b_\beta$ actually span these irreducible representations c. The coefficients in the linear combinations, termed coupling coefficients, are written $\langle ab\alpha\beta \mid abc\gamma\rangle$. Thus

$$h^c_\gamma = \sum_{\alpha\beta} \langle ab\alpha\beta \mid abc\gamma\rangle f^a_\alpha g^b_\beta \tag{2.1}$$

It is evident that if we took a number $N(c)$ which depends on c but not on α, β, or γ then the sets H^c_γ given by

$$H^c_\gamma = \sum_{\alpha\beta} N(c) \langle ab\alpha\beta \mid abc\gamma\rangle f^a_\alpha g^b_\beta$$

would do just as well as the h^c_γ. So the coupling coefficients are not completely determined by the conditions we have imposed upon them so far. We now require that they shall be real and that the transformation (2.1) in $\lambda(a)\lambda(b)$ variables shall be orthogonal. Hence

$$\sum_{\alpha\beta} \langle abc\gamma \mid ab\alpha\beta\rangle \langle ab\alpha\beta \mid abc'\gamma'\rangle = \delta_{cc'}\delta_{\gamma\gamma'}$$
$$\sum_{c\gamma} \langle ab\alpha\beta \mid abc\gamma\rangle \langle abc\gamma \mid ab\alpha'\beta'\rangle = \delta_{\alpha\alpha'}\delta_{\beta\beta'} \qquad (2.2)$$

where the first of the two equations is really true only if c is contained in the direct product ab. To make our equations independent of this requirement, we define the symbol $\delta(a, b, c)$ by the rules $\delta(a, b, c) = 1$ if c is in ab and $\delta(a, b, c) = 0$ otherwise. It is immediately obvious that c is in ab when, and only when, A_1 is in the triple direct product abc. Hence, when $\delta(a, b, c) = 1$, we have also a in bc and b in ac, so

$$\delta(b, c, a) = \delta(c, b, a) = \delta(a, c, b) = \delta(c, a, b) = \delta(b, a, c) = 1$$

Similarly, when $\delta(a, b, c) = 0$. Equations (2.2) are now always true if we introduce the factor $\delta(a, b, c)$ into the right-hand side of the first, but not the second, of the two equations.

If, now, we wish to take an alternative set of coupling coefficients, $N(c)\langle ab\alpha\beta \mid abc\gamma\rangle$, still real and satisfying Eqs. (2.2), we must have $N(c)$ real and $N(c)^2 = 1$ for each c in ab. Hence $N(c) = \pm 1$. Thus, apart from this ambiguity of sign, the numerical values of all the coupling coefficients for the octahedral group are already entirely implicit. We have at our disposal still one independent choice of sign for each ordered trio abc. It is important to recognize this freedom of choice clearly; for example, we may choose the sign separately for each of ET_1T_1, T_1ET_1, and T_1T_1E. In Section 2.5 we shall see how to do so in such a way that the coupling coefficients have a very simple and useful behaviour under permutation of the constituent representations.

Sets of values of the coupling coefficients for the octahedral group are given in references (20) and (26). These values differ from each other and from the values we shall adopt here in certain of their choices of sign. The relationship between the three sets is given in Appendix A in order to make it easy to compare the phases of matrix elements calculated by the methods of this book with those calculated elsewhere.

2.3 REDUCED MATRIX ELEMENTS

Our theoretical method is built upon two foundation stones—the concept of reduced matrix element and the symmetry of the coupling coefficients. The first of these may be derived from Griffith (26), Eq. (8.31), which asserts that if we have functions $|a\alpha\rangle$ and $|a'\alpha'\rangle$ spanning a and a' respectively, together with a set of operators g^b_β spanning b, with a, a', and b irreducible, then

$$\langle a\alpha \mid g^b_\beta \mid a'\alpha'\rangle = K\langle ba'a\alpha \mid ba'\beta\alpha'\rangle \tag{2.3}$$

where K is the same for all $\lambda(a)\lambda(a')\lambda(b)$ choices of α, α', β. Equation (2.3) may be regarded as a factorization of the matrix element into a product of an intrinsic, or scalar, part K and a part $\langle ba'a\alpha \mid ba'\beta\alpha'\rangle$, depending upon which particular components have been selected. Alternatively, we may regard the two parts as those which, respectively, are not and are determined by group-theoretic considerations. Just as with the coupling coefficients, so here there is some leeway in our precise choice of factorization: we could incorporate a numerical factor n with K if we simultaneously incorporated n^{-1} with $\langle ba'a\alpha \mid ba'\beta\alpha'\rangle$, without altering the truth of Eq. (2.3). After making a definite choice about the value of n, we call Kn a reduced matrix element and write it $\langle a \| g^b \| a'\rangle$. Thus Eq. (2.3) becomes

$$\langle a\alpha \mid g^b_\beta \mid a'\alpha'\rangle = n^{-1}\langle a \| g^b \| a'\rangle\langle ba'a\alpha \mid ba'\beta\alpha'\rangle \tag{2.4}$$

where n will be chosen in Section 2.5.

2.4 SYMMETRY OF THE COUPLING COEFFICIENTS

Suppose we take g^b_β to be real functions and, in accord with the plan outlined at the end of Section 2.1, $|a\alpha\rangle$ and $|a'\alpha'\rangle$ also to be real. Then, clearly,

$$\langle a'\alpha' \mid g^b_\beta \mid a\alpha\rangle = \overline{\langle a\alpha \mid g^b_\beta \mid a'\alpha'\rangle} = \langle a\alpha \mid g^b_\beta \mid a'\alpha'\rangle$$

However, because of correspondence to Eq. (2.3), we have for this new matrix element also

$$\langle a'\alpha' \mid g^b_\beta \mid a\alpha\rangle = L\langle baa'\alpha' \mid ba\beta\alpha\rangle$$

for some L independent of α, α', and β. Combining our findings, we deduce

$$K\langle ba'a\alpha \mid ba'\beta\alpha'\rangle = L\langle baa'\alpha' \mid ba\beta\alpha\rangle \tag{2.5}$$

where, of course, K and L are both real. Now we square Eq. (2.5) and sum over α, α', and β, using the first of Eqs. (2.2). We have first

$$K^2 \sum_{\alpha\alpha'\beta} \langle ba'a\alpha \mid ba'\beta\alpha'\rangle^2 = L^2 \sum_{\alpha\alpha'\beta} \langle baa'\alpha' \mid ba\beta\alpha\rangle^2$$

and then $K^2\lambda(a) = L^2\lambda(a')$, provided $\delta(a, a', b) = 1$, which implies

$$K\lambda(a)^{1/2} = \pm L\lambda(a')^{1/2}$$

and

$$\lambda(a)^{-1/2}\langle ba'a\alpha \mid ba'\beta\alpha'\rangle = \pm\lambda(a')^{-1/2}\langle baa'\alpha' \mid ba\beta\alpha\rangle$$

with the sign $\pm$ the same for all α, α', and β. If $\delta(a, a', b) = 0$, there are no coupling coefficients, so we have shown that the coefficients always possess symmetry with respect to exchange of $a\alpha$ with $a'\alpha'$.

There are six possible orders for a, a' and b, and we naturally ask whether the coupling coefficients are simply related to each other for all six orders. We prove that they are by resorting to a simple device. Let f_α^a, g_β^b, $h_{\alpha'}^{a'}$ be three sets of real functions having the behaviour indicated by their suffixes. Also let $\mid\phi\rangle$ and $\mid\psi\rangle$ both span the unit representation A_1 of O. Then both $f_\alpha^a \mid\phi\rangle$ and $f_\alpha^a \mid\psi\rangle$ are evidently sets of kets spanning the irreducible representation a with components α, just as the f_α^a themselves do. A similar result holds for g_β^b and $h_{\alpha'}^{a'}$. According to Eq. (2.3), we now have

$$\langle\phi \mid f_\alpha^a g_\beta^b h_{\alpha'}^{a'} \mid \psi\rangle = K_1\langle ba'a\alpha \mid ba'\beta\alpha'\rangle$$

$$\langle\phi \mid g_\beta^b h_{\alpha'}^{a'} f_\alpha^a \mid \psi\rangle = K_2\langle a'ab\beta \mid a'a\alpha'\alpha\rangle, \text{ etc.}$$

But, clearly, the matrix elements on the left are the same no matter what the order of f_α^a, g_β^b, and $h_{\alpha'}^{a'}$, so

$$K_{11}\langle b_1a_1'a_1\alpha_1 \mid b_1a_1'\beta_1\alpha_1'\rangle = K_1\langle ba'a\alpha \mid ba'\beta\alpha'\rangle$$

for $a_1\alpha_1$, $a_1'\alpha_1'$, $b_1\beta_1$, and any of the six possible permutations of $a\alpha$, $a'\alpha'$, $b\beta$. Hence, just as before, we deduce

$$\lambda(a_1)^{-1/2}\langle b_1a_1'a_1\alpha_1 \mid b_1a_1'\beta_1\alpha_1'\rangle = \pm\lambda(a)^{-1/2}\langle ba'a\alpha \mid ba'\beta\alpha'\rangle \tag{2.6}$$

The sign on the right-hand side of Eq. (2.6) is not necessarily the same for different permutations of $a\alpha$, $a'\alpha'$, $b\beta$.

2.5 V COEFFICIENTS

The V coefficients are defined in terms of the coupling coefficients by the equation

$$V\begin{pmatrix} a & b & c \\ \alpha & \beta & \gamma \end{pmatrix} = \lambda(c)^{-1/2}\langle ab\alpha\beta \mid abc\gamma\rangle \tag{2.7}$$

It is evident from the preceding discussion that the V coefficients are, at most, changed in sign by any permutation of their columns and that in an equation such as

$$V\begin{pmatrix} c & b & a \\ \gamma & \beta & \alpha \end{pmatrix} = \pm V\begin{pmatrix} a & b & c \\ \alpha & \beta & \gamma \end{pmatrix}$$

the sign on the right-hand side is the same for all choices of α, β, and γ. We now recall that we have not specified our decisions of sign for the sets of coupling coefficients. It would be natural to try to do so in such a way that the V become invariant to all permutations of their columns. It will appear shortly, however, that this invariance cannot always be achieved, so we must consider the problem in somewhat more general terms. As each V suffers, at most, a sign change under a permutation, there are two particularly simple kinds of behaviour possible. One is the invariance just mentioned and the other is that V should always be multiplied by the parity of the permutation. We shall say that V is "even" or "odd" according to whether it has the first or the second of these two behaviours. It is not necessary that V should be either even or odd, although we shall find that all our V can be chosen to be one or the other or, sometimes, either. The theoretical method described in this book has the evenness or oddness of the V as an essential part of its structure. It would be perfectly possible, however, to define the V to have a different and more complicated behaviour under permutation and, although it is perhaps unlikely, there might be situations in which such an alternative would be more useful.

Returning to the choice of signs, we first note that if a, b, and c are all different, then each of the six permutations of the columns of V in Eq. (2.7) gives a set of V coefficients which is proportional to an entirely different set of coupling coefficients on the right-hand side. Since we have a separate sign choice for each set of coupling coefficients, we can choose whatever behaviour we wish for V under permutations. In particular, we may choose V to be either even or odd.

With the octahedral group there are two cases of this kind: $A_2T_1T_2$ and ET_1T_2. The first is particularly simple, because

$$\pm\langle A_2T_1T_2\gamma \mid A_2T_{1}\iota\beta\rangle = \pm\langle T_1A_2T_2\gamma \mid T_1A_2\beta\iota\rangle$$
$$= \pm\langle A_2T_2T_1\gamma \mid A_2T_{2}\iota\beta\rangle = \pm\langle T_2A_2T_1\gamma \mid T_2A_2\beta\iota\rangle = \delta_{\beta\gamma}$$

and
$$\pm\langle T_1T_2A_2\iota \mid T_1T_2\alpha\beta\rangle = \pm\langle T_2T_1A_2\iota \mid T_2T_1\alpha\beta\rangle = \frac{1}{\sqrt{3}}\delta_{\alpha\beta}$$

From these we deduce

$$V\begin{pmatrix} A_2 & T_1 & T_2 \\ \iota & \beta & \gamma \end{pmatrix} = \pm\frac{1}{\sqrt{3}}\delta_{\beta\gamma}, \ldots, V\begin{pmatrix} T_2 & T_1 & A_2 \\ \alpha & \beta & \iota \end{pmatrix} = \pm\frac{1}{\sqrt{3}}\delta_{\alpha\beta}$$

By taking the plus sign throughout, V becomes invariant to all permutations of its columns. There are then just three non-zero V for each order of A_2, T_1, T_2, as shown in Table 2.3, and the advantage of using the same symbols for the components of T_1 and T_2 is already beginning to appear.

Table 2.3. Non-zero V coefficients for $A_2T_1T_2$.

A_2	T_1	T_2	V
ι	x	x	$1/\sqrt{3}$
ι	y	y	$1/\sqrt{3}$
ι	z	z	$1/\sqrt{3}$

The V coefficients for ET_1T_2 are easily deduced from the known values of the coupling coefficients (20, 26). All the V that we are deriving for the octahedral group are given in Appendix C.

This is a convenient point at which to remind the reader that it has been usual in ligand-field theory to make an assumption of symmetry for the coupling coefficients which reads

$$\langle bac\gamma \mid ba\beta\alpha\rangle = \langle abc\gamma \mid ab\alpha\beta\rangle \tag{2.8}$$

whenever $a \neq b$ (20, 26). This assumption has to be thrown aside in the present work. However, that is no hardship, for Eq. (2.8) was accepted previously not because it really simplified anything but because there was no reason to assume anything else. We shall see shortly that for $abc = A_2EE$ it is not possible to choose that V be even, and we take V to be odd, which is inconsistent with Eq. (2.8). So we do not assume Eq. (2.8).

Now let two of the representations be the same and the third be A_1. In place of the six independent sets of coupling coefficients that we had for $A_2T_1T_2$, there are now just three, namely

$$\pm\langle A_1bb\gamma \mid A_1b\iota\beta\rangle = \pm\langle bA_1b\gamma \mid bA_1\beta\iota\rangle = \delta_{\beta\gamma}$$

and

$$\pm\langle bbA_1\iota \mid bb\beta\gamma\rangle = \lambda(b)^{-1/2}\delta_{\beta\gamma}$$

The last of these follows easily from the fact that if the set of functions f_β^b transform as b, then the quantity $\psi = \Sigma\,(f_\beta^b)^2$ is left unchanged by all elements of the group simply because each element g operates on f_β^b according to an equation

$$gf_\beta^b = \sum f_\gamma^b M_{\gamma\beta}(g)$$

with $M_{\gamma\beta}$ an orthogonal matrix. So ψ becomes

$$\sum_\beta (gf_\beta^b)^2 = \sum_{\beta\gamma\epsilon} f_\gamma^b f_\epsilon^b M_{\gamma\beta}M_{\epsilon\beta}$$

$$= \sum_{\gamma\epsilon} f_\gamma^b f_\epsilon^b \delta_{\gamma\epsilon} = \sum_\gamma f_\gamma^b f_\gamma^b = \psi$$

We choose the plus sign throughout and then have

$$V\begin{pmatrix} b & b & A_1 \\ \beta & \gamma & \iota \end{pmatrix} = \lambda(b)^{-1/2}\delta_{\beta\gamma} \tag{2.9}$$

Equation (2.9) takes V to be even. However, in this case it is not possible for V to be odd. If it were, we should have

$$V\begin{pmatrix} b & b & A_1 \\ \beta & \gamma & \iota \end{pmatrix} = -V\begin{pmatrix} b & b & A_1 \\ \gamma & \beta & \iota \end{pmatrix}$$

and when we put $\beta = \gamma$, we could deduce $V\begin{pmatrix} b & b & A_1 \\ \beta & \beta & \iota \end{pmatrix} = 0$, which is untrue.

We now discuss $a = b \neq c$ in general. We recall that the direct product of a with itself may be split up into two parts—the symmetrized and antisymmetrized squares $[a^2]$ and (a^2). If we have two sets of functions f^a_α and g^a_α, then $[a^2]$ is spanned by the symmetrized products $f^a_\alpha g^a_\beta + f^a_\beta g^a_\alpha$ and (a^2) by $f^a_\alpha g^a_\beta - f^a_\beta g^a_\alpha$. It follows at once that the coupling coefficients satisfy

$$\langle aac\gamma \mid aa\alpha\beta\rangle = \epsilon\langle aac\gamma \mid aa\beta\alpha\rangle$$

where $\epsilon = 1$ when c is in $[a^2]$ and $\epsilon = -1$ when c is in (a^2). In terms of V coefficients, this last equation reads

$$V\begin{pmatrix} a & a & c \\ \alpha & \beta & \gamma \end{pmatrix} = \epsilon V\begin{pmatrix} a & a & c \\ \beta & \alpha & \gamma \end{pmatrix} \tag{2.10}$$

We are still free to choose the signs of the coupling of ac and ca, and we do so in order that

$$V\begin{pmatrix} a & c & a \\ \beta & \gamma & \alpha \end{pmatrix} = V\begin{pmatrix} c & a & a \\ \gamma & \alpha & \beta \end{pmatrix} = V\begin{pmatrix} a & a & c \\ \alpha & \beta & \gamma \end{pmatrix}$$

It then follows from Eq. (2.10) that

$$V\begin{pmatrix} a & c & a \\ \alpha & \gamma & \beta \end{pmatrix} = V\begin{pmatrix} c & a & a \\ \gamma & \beta & \alpha \end{pmatrix} = V\begin{pmatrix} a & a & c \\ \beta & \alpha & \gamma \end{pmatrix} = \epsilon V\begin{pmatrix} a & a & c \\ \alpha & \beta & \gamma \end{pmatrix}$$

so V is even or odd according to whether $\epsilon = 1$ or -1.

For the octahedral group the symmetrized squares are

$$[A_1^2] = [A_2^2] = A_1, \qquad [E^2] = A_1 + E, \qquad [T_1^2] = [T_2^2] = A_1 + E + T_2$$

so V is chosen even for $A_2^2A_1$, E^2A_1, $T_1^2A_1$, T_1^2E, $T_1^2T_2$, $T_2^2A_1$, and T_2^2E. The non-vanishing antisymmetrized squares are $(E^2) = A_2$, $(T_1^2) = (T_2^2) = T_1$, so V is chosen odd for E^2A_2 and $T_2^2T_1$.

Finally, suppose $a = b = c$. Then each permutation of V multiplies it by ±1, as before. For example,

$$V\begin{pmatrix} a & a & a \\ \alpha & \beta & \gamma \end{pmatrix} = \epsilon V\begin{pmatrix} a & a & a \\ \beta & \gamma & \alpha \end{pmatrix} \tag{2.11}$$

with $\epsilon = \pm 1$. But as well as permuting the columns of V, we can change the names of the labels for the arbitrary components in Eq. (2.11); this change, of course, cannot alter the truth of the equation. So we have also

$$V\begin{pmatrix} a & a & a \\ \beta & \gamma & \alpha \end{pmatrix} = \epsilon V\begin{pmatrix} a & a & a \\ \gamma & \alpha & \beta \end{pmatrix} \tag{2.12}$$

and

$$V\begin{pmatrix} a & a & a \\ \gamma & \alpha & \beta \end{pmatrix} = \epsilon V\begin{pmatrix} a & a & a \\ \alpha & \beta & \gamma \end{pmatrix} \tag{2.13}$$

Combining Eqs. (2.11) through (2.13), we find

$$V\begin{pmatrix} a & a & a \\ \alpha & \beta & \gamma \end{pmatrix} = \epsilon^3 V\begin{pmatrix} a & a & a \\ \alpha & \beta & \gamma \end{pmatrix}$$

for all α, β, γ. So provided $\delta(a, a, a) = 1$, i.e., not all the V are zero, we can deduce $\epsilon^3 = 1$ and hence $\epsilon = 1$. This shows also that

$$V\begin{pmatrix} a & a & a \\ \beta & \alpha & \gamma \end{pmatrix} = V\begin{pmatrix} a & a & a \\ \alpha & \gamma & \beta \end{pmatrix} = V\begin{pmatrix} a & a & a \\ \gamma & \beta & \alpha \end{pmatrix}$$

so V is even or odd depending on whether $\eta = +1$ or -1 in the relation

$$V\begin{pmatrix} a & a & a \\ \beta & \alpha & \gamma \end{pmatrix} = \eta V\begin{pmatrix} a & a & a \\ \alpha & \beta & \gamma \end{pmatrix}$$

It is obvious that, equally, V is even or odd according as a appears in $[a^2]$ or (a^2). For the group O we have V even for A_1^3, E^3 and T_2^3 and odd for T_1^3.

We have shown that all our V can be taken to be even or odd and that we have no choice of which except in the two cases $A_2T_1T_2$ and ET_1T_2. V is now always invariant to even, i.e., circular, permutations. Its behaviour

Table 2.4. Behaviour of V under permutation of columns. $A_2T_1T_2$ is chosen even and ET_1T_2 odd.

Even	A_1^3,	$A_1A_2^2$,	A_1E^2,	$A_1T_1^2$,	$A_1T_2^2$,
	E^3,	ET_1^2,	ET_2^2,	$T_1^2T_2$,	T_2^3,
Odd	A_2E^2,	T_1^3,	$T_1T_2^2$		
Either	$A_2T_1T_2$,	ET_1T_2			

under odd permutations, tabulated in Table 2.4, can be formalized in a simple way by introducing a mathematical device. We define the quantity

$(-1)^a$, where a is a representation symbol, to satisfy the usual sort of rules that it would satisfy if a were a whole number; i.e.,

$$(-1)^{a+b} = (-1)^a(-1)^b = (-1)^{b+a}, \qquad (-1)^a = \pm 1$$

We now try to choose the actual values so that $V\begin{pmatrix} a & b & c \\ \alpha & \beta & \gamma \end{pmatrix}$ shall be multiplied by $(-1)^{a+b+c}$ when its columns undergo any odd permutation. $(-1)^{2a}$ must be $+1$ always, so $V\begin{pmatrix} a & a & c \\ \alpha & \beta & \gamma \end{pmatrix}$ is multiplied by $(-1)^c$. The latter property can be so only if $(-1)^c = +1$ if c occurs in any symmetrized square at all and $(-1)^c = -1$ if c is in any antisymmetrized square. Fortunately, A_1, E, and T_2 occur only in the former and A_2 and T_1 only in the latter. So we define

$$\begin{aligned} (-1)^{A_1} = (-1)^E = (-1)^{T_2} &= +1 \\ (-1)^{A_2} = (-1)^{T_1} &= -1 \end{aligned} \tag{2.14}$$

These equations achieve our object for all the V except those for which a, b, and c are different. However, in such cases we can choose freely whether V is even or odd and we do so as Eq. (2.14) dictates. Thus $A_2T_1T_2$ is taken to be even, because $(-1)^{A_2+T_1+T_2} = +1$; ET_1T_2 is taken to be odd, because $(-1)^{E+T_1+T_2} = -1$.

2.6 THE KRAMERS OPERATOR

We remarked at the end of Section 2.1 that if we took our linear operators to be real as linear operators they would not necessarily be real under Kramers's star operator. It is now time to consider the significance of these two conjugacy operations for the reduced matrix elements.

First we make our definition of the reduced matrix elements precise; we choose n of Eq. (2.4) to be $\lambda(a)^{1/2}(-1)^{a+a'+b}$ so that

$$\langle a\alpha \mid g^b_\beta \mid a'\alpha' \rangle = \langle a \,\|\, g^b \,\|\, a' \rangle \, V\begin{pmatrix} a & a' & b \\ \alpha & \alpha' & \beta \end{pmatrix} \tag{2.15}$$

It is convenient to define $\langle a \,\|\, g^b \,\|\, a' \rangle = 0$ whenever $\delta(a, b, a') = 0$. The order of a, b, and a' in the V symbol will always be taken as in Eq. (2.15). It would not, of course, matter if we took a different even permutation of the columns of V, but an odd permutation would multiply V by $(-1)^{a+a'+b}$.

Now suppose g^b_β satisfies $\overline{g^b_\beta} = g^b_\beta = \eta g^{b*}_\beta$ where $\eta = \pm 1$ is the same for each β. For example, if g^b_β is the total orbital angular momentum L_β, then $\overline{L}_\beta = L_\beta = -L^*_\beta$, but if g^b_β is the radius vector (x, y, z) then $\overline{g^b_\beta} = g^b_\beta = g^{b*}_\beta$. Taking the two complex conjugates of Eq. (2.15) gives first

$$\langle a'\alpha' \mid g^b_\beta \mid a\alpha\rangle = \overline{\langle a\alpha \mid g^b_\beta \mid a'\alpha'\rangle} = \overline{\langle a \parallel g^b \parallel a'\rangle} V \begin{pmatrix} a & a' & b \\ \alpha & \alpha' & \beta \end{pmatrix}$$

from which, using

$$\langle a'\alpha' \mid g^b_\beta \mid a\alpha\rangle = \langle a' \parallel g^b \parallel a\rangle V \begin{pmatrix} a' & a & b \\ \alpha' & \alpha & \beta \end{pmatrix}$$

$$= \langle a' \parallel g^b \parallel a\rangle (-1)^{a+a'+b} V \begin{pmatrix} a & a' & b \\ \alpha & \alpha' & \beta \end{pmatrix}$$

we deduce

$$\langle a' \parallel g^b \parallel a\rangle = (-1)^{a+a'+b} \overline{\langle a \parallel g^b \parallel a'\rangle} \tag{2.16}$$

and, secondly,

$$\overline{\langle a \parallel g^b \parallel a'\rangle} V \begin{pmatrix} a & a' & b \\ \alpha & \alpha' & \beta \end{pmatrix} = \overline{\langle a\alpha \mid g^b_\beta \mid a'\alpha'\rangle}$$

$$= \langle a\alpha \mid g^b_\beta \mid a'\alpha'\rangle^* = \eta \langle a\alpha \mid g^b_\beta \mid a'\alpha'\rangle$$

$$= \eta \langle a \parallel g^b \parallel a'\rangle V \begin{pmatrix} a & a' & b \\ \alpha & \alpha' & \beta \end{pmatrix}$$

whence

$$\overline{\langle a \parallel g^b \parallel a'\rangle} = \eta \langle a \parallel g^b \parallel a'\rangle \tag{2.17}$$

We used the fact that $\bar{\rho} = \rho^*$ for any number ρ. Equation (2.16) shows how to change the order of the functions $|a\alpha\rangle$ and $|a'\alpha'\rangle$ in the reduced matrix element, whereas Eq. (2.17) shows that $\langle a \parallel g^b \parallel a'\rangle$ is real or purely imaginary as $\eta = +1$ or -1, respectively.

It is often necessary to calculate the matrix elements of an operator g^b_β within a single set of states $|a\alpha\rangle$. These matrix elements are all multiples of the single quantity $\langle a \parallel g^b \parallel a\rangle$. But now, putting $a = a'$ into Eqs. (2.16) and (2.17), we have two relationships between $\langle a \parallel g^b \parallel a\rangle$ and its complex conjugate, namely,

$$\overline{\langle a \parallel g^b \parallel a\rangle} = (-1)^b \langle a \parallel g^b \parallel a\rangle \quad \text{and} \quad \overline{\langle a \parallel g^b \parallel a\rangle} = \eta \langle a \parallel g^b \parallel a\rangle$$

Hence, unless $\eta = (-1)^b$, we deduce $\langle a \parallel g^b \parallel a\rangle = 0$. Consequently, the reduced matrix element is necessarily zero for $\eta = +1$ and $b = A_2$ or T_1 and also for $\eta = -1$ and $b = A_1$, E, or T_2. The orbital angular momentum L has $\eta = -1$ and $b = T_1$, so there are no restrictions upon it implied by our present findings. However, we can deduce that all its matrix elements in our present scheme are pure imaginary.

The reader will perhaps notice that our treatment depends critically upon the value of η, and yet this value could be altered by multiplying g^b_β by $i = \sqrt{-1}$. Doing so changes the sign of g^{b*}_β. However, it also changes the sign of $\overline{g^b_\beta}$, and the preceding discussion is essentially unchanged. If we were to allow g^b_β to be a real or purely imaginary linear operator, we would take the definition of η as the ratio of $\overline{g^b_\beta}$ to g^{b*}_β. The really significant feature of g^b_β is whether it behaves in the same way or in opposed ways under the bar and star operation. Finally, it would be possible for g^b_β to behave in a mixed way under bar, star, or both, and then we should naturally consider g^b_β broken up into real and imaginary parts under the two operations and discuss each piece separately.

2.7 ELEMENTARY PROPERTIES OF V COEFFICIENTS

The orthonormality relations expressed by Eqs. (2.2) for the coupling coefficients may be rewritten in terms of the V, giving

$$\begin{gathered}\sum_{\alpha\beta} V\begin{pmatrix} a & b & c \\ \alpha & \beta & \gamma \end{pmatrix} V\begin{pmatrix} a & b & c' \\ \alpha & \beta & \gamma' \end{pmatrix} = \lambda(c)^{-1}\delta_{cc'}\delta_{\gamma\gamma'}\delta(a, b, c) \\ \sum_{c\gamma} \lambda(c) V\begin{pmatrix} a & b & c \\ \alpha & \beta & \gamma \end{pmatrix} V\begin{pmatrix} a & b & c \\ \alpha' & \beta' & \gamma \end{pmatrix} = \delta_{\alpha\alpha'}\delta_{\beta\beta'}\end{gathered} \tag{2.18}$$

Now we put $c = c'$ and $\gamma = \gamma'$ in the first of these and sum over γ to give

$$\sum_{\alpha\beta\gamma} V\begin{pmatrix} a & b & c \\ \alpha & \beta & \gamma \end{pmatrix}^2 = \delta(a, b, c) \tag{2.19}$$

showing the V to satisfy $|V| \leqslant 1$.

Remembering that the product bc does not contain A_1 unless $b = c$, we have the slight generalization of Eq. (2.9):

$$V\begin{pmatrix} b & c & A_1 \\ \beta & \gamma & \iota \end{pmatrix} = \lambda(b)^{-1/2}\delta_{bc}\delta_{\beta\gamma} \tag{2.20}$$

Putting $\gamma = \beta$ and summing this over β, we find

$$\sum_{\beta} V\begin{pmatrix} b & b & A_1 \\ \beta & \beta & \iota \end{pmatrix} = \lambda(b)^{1/2}$$

This result can be obtained alternatively in a more general form by putting $c' = A_1$ and $a = b$ in the first of Eqs. (2.18), whence

$$\sum_{\beta} V\begin{pmatrix} b & b & c \\ \beta & \beta & \gamma \end{pmatrix} = \lambda(b)^{1/2}\delta_{cA_1} \tag{2.21}$$

When a, b, and c are all T representations, there are simple explicit formulae for the V. They depend upon the so-called alternating tensor $\epsilon_{\alpha\beta\gamma}$. This is defined to be zero if any two of α, β, or γ are equal, to be $+1$ if

$\alpha\beta\gamma$ is an even permutation of xyz and -1 if it is an odd permutation. $\epsilon_{\alpha\beta\gamma}$ satisfies a rather famous relation, namely,

$$\sum_{\alpha} \epsilon_{\alpha\beta\gamma}\epsilon_{\alpha\beta'\gamma'} = \delta_{\beta\beta'}\delta_{\gamma\gamma'} - \delta_{\beta\gamma'}\delta_{\gamma\beta'} \tag{2.22}$$

Equation (2.22), which will be of some use to us later, may be proved by an enumeration of cases.

It is rather easy to verify by inspection of the coupling coefficients that V can be chosen to satisfy the equations

$$\begin{aligned} V\begin{pmatrix} T_1 & T_1 & T_1 \\ \alpha & \beta & \gamma \end{pmatrix} &= V\begin{pmatrix} T_1 & T_2 & T_2 \\ \alpha & \beta & \gamma \end{pmatrix} = -\frac{1}{\sqrt{6}}\epsilon_{\alpha\beta\gamma} \\ V\begin{pmatrix} T_1 & T_1 & T_2 \\ \alpha & \beta & \gamma \end{pmatrix} &= V\begin{pmatrix} T_2 & T_2 & T_2 \\ \alpha & \beta & \gamma \end{pmatrix} = -\frac{1}{\sqrt{6}}|\epsilon_{\alpha\beta\gamma}| \end{aligned} \tag{2.23}$$

and we choose V to satisfy Eqs. (2.23). The desired behaviour of the V under permutation of their columns is, of course, consistent with Eqs. (2.23) and the properties of $\epsilon_{\alpha\beta\gamma}$.

2.8 ORTHOGONAL TRANSFORMATIONS OF COMPONENTS

We have chosen definite sets of matrices representing each of our irreducible representations. However, we may sometimes wish to use different sets. Then the new components of functions spanning an irreducible representation, for example, a, will be related to the old by an orthonormal transformation

$$f^a_{\alpha\dagger} = \sum_{\alpha} A_{\alpha\dagger\alpha} f^a_\alpha \tag{2.24}$$

For a given a, $A_{\alpha\dagger\alpha}$ is the same no matter what particular functions f^a_α are being considered. So there is one matrix $A_{\alpha\dagger\alpha}$ for each of the five irreducible representations. We still take $f^a_{\alpha\dagger}$ to be real, so $A_{\alpha\dagger\alpha}$ is actually an orthogonal matrix.

The law of transformation for the V is determined by the obvious requirement that the reduced matrix elements should be left invariant. The factor $(-1)^{a+b+c}$ being also invariant, we deduce that $V\begin{pmatrix} a & b & c \\ \alpha & \beta & \gamma \end{pmatrix}$ must have the same transformation law as the matrix element $\langle a\alpha \mid g^b_\beta \mid c\gamma\rangle$; i.e.,

$$V\begin{pmatrix} a & b & c \\ \alpha\dagger & \beta\dagger & \gamma\dagger \end{pmatrix} = \sum_{\alpha\beta\gamma} A_{\alpha\dagger\alpha} B_{\beta\dagger\beta} C_{\gamma\dagger\gamma} V\begin{pmatrix} a & b & c \\ \alpha & \beta & \gamma \end{pmatrix} \tag{2.25}$$

We then verify by direct expansion that the reduced matrix element is the same in the new as in the old component system. Thus

$$\begin{aligned}\langle a\alpha\dagger \mid g^b_{\beta\dagger} \mid c\gamma\dagger\rangle &= \sum_{\alpha\beta\gamma} A_{\alpha\dagger\alpha}B_{\beta\dagger\beta}C_{\gamma\dagger\gamma}\langle a\alpha \mid g^b_\beta \mid c\gamma\rangle \\ &= \langle a \| g^b \| c\rangle \sum A_{\alpha\dagger\alpha}B_{\beta\dagger\beta}C_{\gamma\dagger\gamma} V\begin{pmatrix} a & c & b \\ \alpha & \gamma & \beta\end{pmatrix} \qquad (2.26) \\ &= \langle a \| g^b \| c\rangle V\begin{pmatrix} a & c & b \\ \alpha\dagger & \gamma\dagger & \beta\dagger\end{pmatrix}\end{aligned}$$

If we use the corresponding properties of $V\begin{pmatrix} a & b & c \\ \alpha & \beta & \gamma\end{pmatrix}$, it also follows at once from Eq. (2.25) that $V\begin{pmatrix} a & b & c \\ \alpha\dagger & \beta\dagger & \gamma\dagger\end{pmatrix}$ is also invariant to even permutations of its columns and is multiplied by $(-1)^{a+b+c}$ by odd permutations.

A transformation of particular interest is that to a trigonal component system. If the trigonal axis is the (1, 1, 1) axis, we may write

$$\begin{aligned} f_\rho &= -\frac{1}{\sqrt{6}}f_x - \frac{1}{\sqrt{6}}f_y + \frac{2}{\sqrt{6}}f_z \\ f_\sigma &= \frac{1}{\sqrt{2}}f_x - \frac{1}{\sqrt{2}}f_y \qquad (2.27) \\ f_\tau &= \frac{1}{\sqrt{3}}f_x + \frac{1}{\sqrt{3}}f_y + \frac{1}{\sqrt{3}}f_z \end{aligned}$$

for both T_1 and T_2. It is convenient to take the same component system for A_1, A_2, and E that we have used heretofore. The trigonal system is useful in problems where there is a perturbation having full trigonal symmetry about the (1, 1, 1) axis, for then the pair (f_ρ, f_σ) and the single function f_τ each span irreducible representations of the dihedral group D_3 about the (1, 1, 1) axis. The E representation of the octahedral group remains irreducible under D_3 and spans the same representation of the latter group as do (f_ρ, f_σ). Further, the functions (θ, ϵ) and (f_ρ, f_σ) give rise to this representation with identically the same matrices. The V coefficients for the trigonal system can be calculated from Eq. (2.25); they are given in Appendix C.

2.9 COMPLEX TRANSFORMATIONS

In discussing magnetic properties it is often desirable to use complex functions. Using complex functions introduces an extra complication into the definition of V coefficients, because in a matrix element $\langle a\alpha\dagger \mid g^b_{\beta\dagger} \mid c\gamma\dagger\rangle$ the bras $\langle a\alpha\dagger|$ no longer necessarily transform in the same way as do the kets $|a\alpha\dagger\rangle$ under group elements. However, our complex functions will be orthonormal transforms of a real set of functions according to Eq. (2.24), so we can still expand any matrix element as we did in deriving Eq. (2.26), to give

$$\langle a\alpha\dagger \mid g^b_{\beta\dagger} \mid c\gamma\dagger\rangle = \langle a \,\|\, g^b \,\|\, c\rangle \sum_{\alpha\beta\gamma} \overline{A}_{\alpha\dagger\alpha} B_{\beta\dagger\beta} C_{\gamma\dagger\gamma} V\begin{pmatrix} a & c & b \\ \alpha & \gamma & \beta \end{pmatrix} \tag{2.28}$$

We have merely to tabulate the sums occurring in Eq. (2.28) for all values of a, b, c, $\alpha\dagger$, $\beta\dagger$, and $\gamma\dagger$ to enable us to calculate easily all matrix elements in the complex scheme in terms of the reduced matrix elements belonging to the real scheme. These sums do not, however, generally have a simple behaviour under permutation of components; it is fortunate, therefore, that in the cases of greatest practical interest we can still define V coefficients which do behave symmetrically. The trick is to use transformations with numerical suffixes $\alpha\dagger$, $\beta\dagger$, $\gamma\dagger$ and with the property that the matrix $A_{\alpha\dagger\alpha}$ is simply related to $A_{-\alpha\dagger\alpha}$.

For calculations in which a magnetic field is applied along the OZ axis, we keep the usual (θ, ϵ) component system for E and, of course, leave A_1 and A_2 unaltered. For both T_1 and T_2 we write

$$\begin{aligned} f_1 &= -\frac{i}{\sqrt{2}} f_x + \frac{1}{\sqrt{2}} f_y \\ f_0 &= i f_z \\ f_{-1} &= \frac{i}{\sqrt{2}} f_x + \frac{1}{\sqrt{2}} f_y \end{aligned} \tag{2.29}$$

It is easy to verify that the matrix $A_{\alpha\dagger\alpha}$ for Eqs. (2.29) satisfies

$$\overline{A}_{\alpha\dagger\alpha} = (-1)^{1+\alpha\dagger} A_{-\alpha\dagger,\alpha} \tag{2.30}$$

In the trigonal component system with a magnetic field along the (1, 1, 1) axis, we would naturally consider the analogous transformations

$$\begin{aligned} f_1 &= -\frac{i}{\sqrt{2}} f_\rho + \frac{1}{\sqrt{2}} f_\sigma \\ f_0 &= i f_\tau \\ f_{-1} &= \frac{i}{\sqrt{2}} f_\rho + \frac{1}{\sqrt{2}} f_\sigma \end{aligned} \tag{2.31}$$

for T_1 and T_2, which also satisfy Eq. (2.30), and

$$\begin{aligned} g_1 &= -\frac{i}{\sqrt{2}} g_\theta + \frac{1}{\sqrt{2}} g_\epsilon \\ g_{-1} &= \frac{i}{\sqrt{2}} g_\theta + \frac{1}{\sqrt{2}} g_\epsilon \end{aligned} \tag{2.32}$$

for E, which satisfy $\overline{A}_{\alpha\dagger\alpha} = A_{-\alpha\dagger\alpha}$.

It is convenient to define a symbol $[-1]^{\xi}$ where $[-1]^{\xi} = 1$ if ξ is a representation or a component referring to A_1, A_2, or E. Thus we even require $[-1]^1 = 1$ when the superfix 1 refers to an E representation. However, if $\xi = T_1$ or T_2, we put $[-1]^{\xi} = -1$; if ξ is a component of T_1 or T_2, we put $[-1]^{\xi} = (-1)^{\xi}$. Then in both the complex systems we have the formula

$$\overline{A}_{\alpha\dagger\alpha} = [-1]^{\alpha+\alpha\dagger} A_{-\alpha\dagger\alpha} \tag{2.33}$$

for each of the five irreducible representations. We interpret $-\alpha\dagger$ as $\alpha\dagger$ when $\alpha\dagger = \iota$, θ, or ϵ.

We are now in a position to define V in our two complex component systems according to Eq. (2.25) whereupon it follows immediately that V retains its usual behaviour under permutation of its columns. With this definition, Eq. (2.28) simplifies to

$$\begin{aligned}\langle a\alpha\dagger \mid g^b_{\beta\dagger} \mid c\gamma\dagger\rangle &= \langle a \| g^b \| c\rangle \sum_{\alpha\beta\gamma} [-1]^{a+\alpha\dagger} A_{-\alpha\dagger\alpha} B_{\beta\dagger\beta} C_{\gamma\dagger\gamma} V\begin{pmatrix} a & c & b \\ \alpha & \gamma & \beta \end{pmatrix} \\ &= [-1]^{a+\alpha\dagger} \langle a \| g^b \| c\rangle V\begin{pmatrix} a & c & b \\ -\alpha\dagger & \gamma\dagger & \beta\dagger \end{pmatrix}\end{aligned} \tag{2.34}$$

Finally, for the complex tetragonal component system, the V are real; hence,

$$\begin{aligned}V\begin{pmatrix} a & b & c \\ \alpha\dagger & \beta\dagger & \gamma\dagger \end{pmatrix} &= \overline{V\begin{pmatrix} a & b & c \\ \alpha\dagger & \beta\dagger & \gamma\dagger \end{pmatrix}} \\ &= \sum_{\alpha\dagger\beta\dagger\gamma\dagger} \overline{A}_{\alpha\dagger\alpha} \overline{B}_{\beta\dagger\beta} \overline{C}_{\gamma\dagger\gamma} V\begin{pmatrix} a & b & c \\ \alpha & \beta & \gamma \end{pmatrix} \\ &= [-1]^{a+\alpha\dagger+b+\beta\dagger+c+\gamma\dagger} V\begin{pmatrix} a & b & c \\ -\alpha\dagger & -\beta\dagger & -\gamma\dagger \end{pmatrix}\end{aligned} \tag{2.35}$$

where we use both Eq. (2.25) and Eq. (2.33).

The quantities $V\begin{pmatrix} a & b & c \\ \alpha\dagger & \beta\dagger & \gamma\dagger \end{pmatrix}$ for our two complex systems are tabulated in Appendix C. If we use them and Eq. (2.34), any matrix element in either complex system may be immediately expressed in terms of the $\langle a \| g^b \| c\rangle$ which were defined with respect to real component systems.

2.10 EXAMPLES OF REDUCED MATRIX ELEMENTS

We now consider briefly some examples which are basic to our later calculations in n-electron systems. Suppose, first, that we have a number μ. Then

$$\langle a\alpha \mid \mu \mid a'\alpha'\rangle = \mu\delta_{aa'}\delta_{\alpha\alpha'}$$

on the one hand, and, in terms of the reduced matrix element of μ,

$$\langle a\alpha \mid \mu \mid a'\alpha' \rangle = \langle a \parallel \mu \parallel a' \rangle V \begin{pmatrix} a & a' & A_1 \\ \alpha & \alpha' & \iota \end{pmatrix} = \lambda(a)^{-1/2}\delta_{aa'}\delta_{\alpha\alpha'}\langle a \parallel \mu \parallel a' \rangle$$

by Eqs. (2.15) and (2.20). We have A_1 in the V coefficient because, obviously, μ is left invariant by all group elements. So

$$\langle a \parallel \mu \parallel a' \rangle = \lambda(a)^{1/2}\mu\delta_{aa'} \tag{2.36}$$

for any number μ.

Next comes the orbital angular momentum vector **l** for a one-electron system. This transforms as T_1, and it is easiest to obtain its reduced matrix elements by working with the complex tetragonal component system. Then Eqs. (2.29) applied to **l** may be rewritten

$$l_0 = il_z, \qquad l_{\pm 1} = \mp \frac{i}{\sqrt{2}} l^{\pm} \tag{2.37}$$

in terms of l_z and the shift operators.

Equations (2.29) were chosen so that if (f_x, f_y, f_z) is a set of three atomic p orbitals (p_x, p_y, p_z) then the derived functions $p_m (m = 1, 0, -1)$ satisfy*

$$l_z p_m = m p_m, \qquad l^{\pm} p_m = \{(1 \mp m)(2 \pm m)\}^{1/2} p_{m\pm 1} \tag{2.38}$$

In other words, the p_m are a set of p orbitals $|1m\rangle$ quantized with respect to l_z and connected in phase according to the usual convention. It follows from Eq. (2.38) that

$$\langle p_1 \mid l_0 \mid p_1 \rangle = i$$

and from Eq. (2.34) that

$$\langle p_1 \mid l_0 \mid p_1 \rangle = \langle p \parallel l \parallel p \rangle V \begin{pmatrix} T_1 & T_1 & T_1 \\ -1 & 0 & 1 \end{pmatrix} = \frac{1}{\sqrt{6}} \langle p \parallel l \parallel p \rangle$$

whence

$$\langle p \parallel l \parallel p \rangle = i\sqrt{6}$$

Using this reduced matrix element means that we can immediately write down the complete matrix of **l** for p orbitals in any component system for which we have tabulated V symbols. For example,

$$\langle p_x \mid l_y \mid p_z \rangle = \langle p \parallel l \parallel p \rangle V \begin{pmatrix} T_1 & T_1 & T_1 \\ x & z & y \end{pmatrix} = i$$

$$\langle p_0 \mid l^- \mid p_1 \rangle = -i\sqrt{2}\,\langle p_0 \mid l_{-1} \mid p_1 \rangle$$

$$= -i\sqrt{2}\,\langle p \parallel l \parallel p \rangle V \begin{pmatrix} T_1 & T_1 & T_1 \\ 0 & 1 & -1 \end{pmatrix} = \sqrt{2}$$

and so forth.

* Throughout the book, angular momenta are measured in units of $\hbar$.

The reduced matrix elements of $\mathbf{l}$ within and between the irreducible representations of the octahedral group spanned by other sets of atomic orbitals are easily derived in a similar way once one knows the numerical details of the breakdown of these sets into irreducible representations. An interesting example here is furnished by the t_2 representation of d orbitals. Writing $|2m\rangle$ as the d orbitals quantized with respect to l_z and correctly connected in phase, we have [ref. (26), Eq. 9.20]

$$|dt_21\rangle = |2, -1\rangle$$

whence

$$\langle dt_21 \mid l_0 \mid dt_21\rangle = -i$$

and

$$\langle dt_2 \,\|\, l \,\|\, dt_2\rangle = -i\sqrt{6}$$

In each of our component systems we have arranged that

$$V\begin{pmatrix} T_1 & T_1 & T_1 \\ \alpha & \gamma & \beta \end{pmatrix} = V\begin{pmatrix} T_2 & T_1 & T_2 \\ \alpha & \gamma & \beta \end{pmatrix} \tag{2.39}$$

for all α, β, and γ. So we can deduce

$$\langle dt_2\alpha \mid l_\beta \mid dt_2\gamma\rangle = -\langle p_\alpha \mid l_\beta \mid p_\gamma\rangle \tag{2.40}$$

This discussion gives a formal proof of the well-known result of Abragam and Pryce [(18); see also ref. (26)] that the matrix of $\mathbf{l}$ within dt_2 orbitals is the exact negative of its matrix within p orbitals. In Appendix B the relation shown in Eq. (2.39) is seen to be not entirely accidental, which gives a somewhat deeper significance to the proportionality in Eq. (2.40). Note that it is obvious also that the matrix of $\mathbf{l}$ or $\mathbf{L}$ within any set, either of T_1 or of T_2 functions, will be proportional to the matrix of $\mathbf{l}$ within p orbitals.

The reduced matrix elements of $\mathbf{l}$ for p, d, and f orbitals are given in Table 2.5. The reader may usefully ponder why the matrices in that table are neither all symmetric nor all skew-symmetric.

Table 2.5. Reduced matrix elements of $\mathbf{l}$ for p, d, and f orbitals.

p	T_1
T_1	$i\sqrt{6}$

d	E	T_2
E	0	$2i\sqrt{3}$
T_2	$2i\sqrt{3}$	$-i\sqrt{6}$

f	A_2	T_1	T_2
A_2	0	0	$2i\sqrt{3}$
T_1	0	$-\frac{3}{2}i\sqrt{6}$	$-\frac{3}{2}i\sqrt{10}$
T_2	$-2i\sqrt{3}$	$\frac{3}{2}i\sqrt{10}$	$\frac{1}{2}i\sqrt{6}$

Spin in n-electron systems is discussed for general n later in the book. However, for $n = 2$ our present approach will suffice. In a two-electron system we may use the total spin S and its M_s value in the classification of a basic set of states and have only the two possible spins $S = 0$ or 1. Hence, under the octahedral group, the spin functions transform as A_1 or T_1. Furthermore, the total states for the system are now simple products

$$\psi = f(\mathbf{r}_1, \mathbf{r}_2)\chi(SM_s) \tag{2.41}$$

of a space function f with a spin function χ. The arguments 1 and 2 number the electrons. Suitable functions χ are

$$\begin{aligned} \chi(0\,0) &= \frac{1}{\sqrt{2}}\alpha(1)\beta(2) - \frac{1}{\sqrt{2}}\alpha(2)\beta(1) \\ \chi(1\,1) &= \alpha(1)\alpha(2) \\ \chi(1\,0) &= \frac{1}{\sqrt{2}}\alpha(1)\beta(2) + \frac{1}{\sqrt{2}}\alpha(2)\beta(1) \\ \chi(1\,-1) &= \beta(1)\beta(2) \end{aligned} \tag{2.42}$$

in terms of the spin functions $\alpha(m_s = +\frac{1}{2})$ and $\beta(m_s = -\frac{1}{2})$ for a single electron. Because of the Pauli principle, ψ must be antisymmetric to interchange of the electrons. $\chi(0\,0)$ is also antisymmetric, but $\chi(1\,M_s)$ are all symmetric. Hence f is symmetric for $S = 0$ and antisymmetric for $S = 1$.

The behaviour of the χ under the octahedral group is correctly represented in the equations

$$\begin{aligned} \chi_\iota^A &= \chi(0\,0) \\ \chi_m^{T_1} &= \chi(1\,m) \end{aligned} \tag{2.43}$$

paralleling the behaviour of the p functions that we have already discussed. Writing $\mathbf{S} = \mathbf{s}(1) + \mathbf{s}(2)$, we can immediately write down the reduced matrix elements of $\mathbf{S}$. We can also obtain the reduced elements of $\mathbf{s}(1)$ and $\mathbf{s}(2)$ separately, using χ given in Eqs. (2.42) and V of Appendix C. These latter elements are

$$\begin{aligned} \langle 0 \| s(1) \| 0\rangle &= \langle 0 \| s(2) \| 0\rangle = 0 \\ \langle 1 \| s(1) \| 1\rangle &= \langle 1 \| s(2) \| 1\rangle = \tfrac{1}{2}i\sqrt{6} \\ \langle 0 \| s(1) \| 1\rangle &= -\langle 0 \| s(2) \| 1\rangle = -\tfrac{1}{2}i\sqrt{3} \\ \langle 1 \| s(1) \| 0\rangle &= -\langle 1 \| s(2) \| 0\rangle = \tfrac{1}{2}i\sqrt{3} \end{aligned} \tag{2.44}$$

where 0 and 1 in the kets and bras refer to $S = 0$ and $S = 1$, respectively. It will be useful for the discussion in Chapter 6 of spin-orbit coupling in two-electron systems to note that it follows from Eqs. (2.44) that the formula

$$\langle S \| s(1) \| S'\rangle = \tfrac{1}{2}(-1)^{S+1}i\sqrt{3}\,(S + S')^{1/2} \tag{2.45}$$

is true for our spin functions.

CHAPTER 3

V Coefficients for Other Symmetry Groups

3.1 GENERAL DISCUSSION

The tensor method is really valuable only for symmetry groups which have at least one irreducible representation of degree greater than 1. Otherwise, it coincides with the procedure for determining selection rules on a matrix element $\langle a\alpha \mid g^b_\beta \mid a'\alpha' \rangle$ by asking whether the direct product $a'b$ contains a or not. A group is of the former kind when, and only when, it is non-commutative. Hence, in developing a theory for molecular symmetry groups or point symmetry groups in lattices, we are potentially interested in all non-commutative subgroups of the three-dimensional orthogonal group (the group of all rotations, reflections, inversion etc., which leave a chosen point of space fixed). It is perfectly reasonable also to contemplate application to space groups, but this application does not appear to have been made yet, and we do not consider it here.

Restricting our attention at first to pure rotation groups, we must consider the series of dihedral groups D_n with $n \geqslant 3$ or $n = \infty$ and also the three groups T, O, and K. The group T is the rotation group of a regular tetrahedron and has 12 elements, O is the octahedral group with 24 elements, and K is the icosahedral group with 60 elements.

It is well known [see Murnaghan (12), pp. 342–4] that a subgroup of the orthogonal group is necessarily either identical with or isomorphic with one of the pure rotation subgroups, or, alternatively, is the direct product of a pure rotation subgroup with the group $(1, P)$ consisting of the unit element and inversion. Hence, providing we analyse the relationship between the theory for a direct product group and that for its two constituents, we need only consider the non-commutative pure rotation groups enumerated in the preceding paragraph.

3.2 DIRECT PRODUCT GROUPS

If we say that a group G is the direct product of two of its subgroups H and K, we mean that if h is any element of H and k of K, then $kh = hk$, and the totality of elements hk gives all the elements of G. Further, H and K have no element in common except the unit element, whence $h_1k_1 = h_2k_2$ when, and only when, $h_1 = h_2$ and $k_1 = k_2$ [see, e.g., (26), pp. 139–140]. Finally, if the orders of G, H, and K are, respectively, π_G, π_H, and π_K, then

$$\pi_G = \pi_H \pi_K \tag{3.1}$$

Now suppose we have an irreducible representation of the group H:

$$hf_i = \sum_{p=1}^{m} a_{pi} f_p \tag{3.2}$$

with the functions f_i as a basis, and character

$$\chi_H(h) = \sum_{p=1}^{m} a_{pp} \tag{3.3}$$

Then the character satisfies the necessary and sufficient condition for irreducibility:

$$\sum_h |\chi_H(h)|^2 = \pi_H \tag{3.4}$$

Similarly, for an irreducible representation of K, we may write

$$kg_j = \sum_{q=1}^{n} b_{qj}g_q$$

$$\chi_K(k) = \sum_{q=1}^{n} b_{qq} \tag{3.5}$$

$$\sum_k |\chi_K(k)|^2 = \pi_K$$

Then the set of mn functions f_ig_j gives a representation of G according to the rule

$$hkf_ig_j = \sum_{p,q} a_{pi}b_{qj}f_pg_q$$

This representation has the character

$$\chi_G(hk) = \sum_{p,q} a_{pp}b_{qq} = \chi_H(h)\chi_K(k)$$

Hence

$$\sum_{g \text{ in } G} |\chi_G(g)|^2 = \sum_{h,k} |\chi_G(hk)|^2$$

$$= \sum_h |\chi_H(h)|^2 \sum_k |\chi_K(k)|^2 = \pi_H\pi_K = \pi_G$$

So our new representation is also irreducible.

Let the irreducible representations of H and K be labelled A_i and B_j, respectively. Then each pair A_i, B_j gives an irreducible representation C_{ij} of G by the above procedure. These are clearly pairwise inequivalent. Further, there are no other inequivalent irreducible representations of G, because if λ_{Hi}, λ_{Kj} are the degrees of A_i and B_j respectively, then

$$\sum_i \lambda_{Hi}^2 = \pi_H, \qquad \sum_j \lambda_{Kj}^2 = \pi_K$$

whence $$\sum \lambda(C_{ij})^2 = \sum_i \lambda_{Hi}^2 \sum_j \lambda_{Kj}^2 = \pi_H\pi_K = \pi_G$$

Consequently, we may write any irreducible representation of G as a formal product A_iB_j. To avoid confusion, it is desirable to choose a definite order for the pair of groups H and K, so we always write A_iB_j and never B_jA_i. If H is a pure rotation group and K has the two elements 1, P, with P the inversion, then we normally write the irreducible representations of K as g and u and put A_{ig} and A_{iu} in place of A_ig and A_iu. We consider other applications of the ideas of this section at the end of Section 9.4 and in Appendix D; hence we prefer the more general formulation given above.

A set of functions transforming as the representation aa' of G may now be written $\phi^{aa'}_{\alpha\alpha'}$. A particular set would be $\phi^{aa'}_{\alpha\alpha'} = f^a_\alpha g^{a'}_{\alpha'}$ with f^a_α left invariant by all elements of K and $g^{a'}_{\alpha'}$ by all elements of H, but it is not necessary that $\phi^{aa'}_{\alpha\alpha'}$ should be factorizable in this manner. Finally, it is slightly tedious but straightforward to verify that, if V coefficients and reduced matrix elements are defined for H and K separately, we may define them for G by the rules

$$V\begin{pmatrix} aa', & bb', & cc' \\ \alpha\alpha', & \beta\beta', & \gamma\gamma' \end{pmatrix} = V\begin{pmatrix} a & b & c \\ \alpha & \beta & \gamma \end{pmatrix} V\begin{pmatrix} a' & b' & c' \\ \alpha' & \beta' & \gamma' \end{pmatrix} \tag{3.6}$$

$$\langle \phi^{aa'}_{\alpha\alpha'} \mid \chi^{bb'}_{\beta\beta'} \mid \psi^{cc'}_{\gamma\gamma'} \rangle = \langle \phi^{aa'} \| \chi^{bb'} \| \psi^{cc'} \rangle V\begin{pmatrix} aa', & cc', & bb' \\ \alpha\alpha', & \gamma\gamma', & \beta\beta' \end{pmatrix} \tag{3.7}$$

It is convenient now to anticipate and to say that we are going to be concerned largely in our tensor method with certain sums of products of V coefficients. Equation (3.6) will show that such sums for a direct product $H \times K$ are simply the products of the corresponding quantities for H and K separately. Thus, for the W coefficients defined at the beginning of Chapter 4, we shall have

$$W\begin{pmatrix} aa', & bb', & cc' \\ dd', & ee', & ff' \end{pmatrix} = W\begin{pmatrix} a & b & c \\ d & e & f \end{pmatrix} W\begin{pmatrix} a' & b' & c' \\ d' & e' & f' \end{pmatrix} \tag{3.8}$$

and we shall have a similar equation for the X coefficients defined in Chapter 8. Consequently, we need define and tabulate our coefficients only for those groups which are not expressible as direct products of pairs of their proper subgroups. We now proceed to consider how to do this.

3.3 THE GROUPS T, O, AND K

The rotation group of a regular tetrahedron, T, has a pair of representations having partially complex characters. Therefore, the sort of difficulty mentioned in Section 2.9 would inevitably be present. It is perfectly straightforward to treat such a case, but as we may usually equally well use the larger group T_d, which has only real characters, we shall discuss classifications with respect only to the latter.

The group T_d is isomorphic with O, so it can be taken to have the same V coefficients. Consequently, all purely group-theoretic results which we obtain for O are equally applicable to T_d. The reader who is interested in tetrahedral symmetry may, in most places in the book, simply replace O with T_d. However, he must remember that a particular set of functions which transforms as a representation of O may transform differently under T_d. For example, the functions (x, y, z) transform as T_1 under O but as T_2 under T_d.

Should the Hamiltonian $\mathcal{H}$ for the system really have the lower symmetry T, we can still classify with respect to T_d, but we shall then have the situation that $\mathcal{H}$ does not necessarily belong to the unit representation of the classifying group. An examination of the character tables of T and T_d shows that $\mathcal{H}$ may be broken into two parts:

$$\mathcal{H} = \mathcal{H}(A_1) + \mathcal{H}(A_2)$$

where $\mathcal{H}(A_1)$ belongs to the unit representation of T_d and $\mathcal{H}(A_2)$ to A_2. This kind of situation also occurs in the theory for spherical symmetry, in which Racah (16) showed that groups which did not leave the Hamiltonian invariant were nevertheless very useful.

We have already discussed O. The product of pairs of irreducible representations of the icosahedral group K sometimes contains representations repeated twice.* So some trios abc would have to have two independent sets [ref. (26), Eq. 8.32, and ref. (27)] of V coefficients associated with them, and Eq. (2.15) would become

$$\langle a\alpha \mid g^b_\beta \mid a'\alpha' \rangle = \langle a \,||\, g^b \,||\, a' \rangle_1 V_1 \begin{pmatrix} a & a' & b \\ \alpha & \alpha' & \beta \end{pmatrix} + \langle a \,||\, g^b \,||\, a' \rangle_2 V_2 \begin{pmatrix} a & a' & b \\ \alpha & \alpha' & \beta \end{pmatrix} \tag{3.9}$$

K is of rather rare occurrence, and because of this extra complexity we do not investigate the necessary generalization of our theory here. We shall not discuss K again.

3.4 DIHEDRAL GROUPS, $D_n(n > 2)$, AND D_∞

These groups are most conveniently treated all at once, and we include those with n twice an odd integer, even though they are expressible as direct products. They all possess two irreducible representations of degree 1, written A_1 and A_2, which are invariant to rotation about the dihedral axis. D_n, for n even and finite, has two further representations of degree 1, which are written B_1 and B_2. These change sign under a rotation of $2\pi/n$ about the dihedral axis. The remaining irreducible representations are of degree 2; we write a typical representation E_m. We take the dihedral axis to be the Z axis and suppose the rotation $C = C_2^x$ of 180 deg. around the x axis to be in the group. Then, if (r, θ, ϕ) are the usual spherical polars, E_m is spanned by $\cos m\phi$ and $\sin m\phi$. We take these functions to define our standard real component system for E_m, and write the first component c and the second s.

* This situation would occur if we were to define V for two-valued representations of O; see also Section 9.7 and Appendix F.

The complete set of inequivalent irreducible representations of degree 2 is obtained by taking all positive integral values of m for D_∞ and those which also satisfy $2m < n$ for D_n.

As an example, consider the behaviour of (x, y, z) under D_3. z transforms as A_2, whereas x and y are, respectively, the c and s components of E (E_1 is usually written E for D_3 and D_4, because it is the only irreducible representation of degree 2).

V coefficients are very easily calculated by using relations like

$$\cos(m\phi_1 + p\phi_2) = \cos m\phi_1 \cos p\phi_2 - \sin m\phi_1 \sin p\phi_2$$

When the three representations in V are all of degree 2, two cases arise. Let the representations be E_m, E_p, E_q. Then either

$$n = m + p + q$$

for the finite group D_n, or one of m, p, q is the sum of the other two (D_n finite or infinite). Corresponding V coefficients are tabulated in Table 3.1.

Table 3.1. V coefficients for three representations, each of degree 2, of a dihedral group D_n.

$n = m + p + q$				$m = p + q$			
E_m	E_p	E_q	V	E_m	E_p	E_q	V
c	c	c	$-\frac{1}{2}$	c	c	c	$\frac{1}{2}$
c	s	s	$\frac{1}{2}$	c	s	s	$-\frac{1}{2}$
s	c	s	$\frac{1}{2}$	s	c	s	$\frac{1}{2}$
s	s	c	$\frac{1}{2}$	s	s	c	$\frac{1}{2}$

The antisymmetrized square of each irreducible representation of degree 2 is A_2, so we define $(-1)^a$ to be -1 when $a = A_2$ and 1 otherwise, and we require $V\begin{pmatrix} a & b & c \\ \alpha & \beta & \gamma \end{pmatrix}$ to be invariant to even permutation of its columns but multiplied by $(-1)^{a+b+c}$ on odd permutation. Consequently, the V of Table 3.1 are invariant to all permutations.

The remaining V with E representations in them are given in Table 3.2. Finally, when V contains only representations of degree 1, we define V for abc to be 1 except for

$$V\begin{pmatrix} B_1 & A_2 & B_2 \\ \iota & \iota & \iota \end{pmatrix} = V\begin{pmatrix} A_2 & B_2 & B_1 \\ \iota & \iota & \iota \end{pmatrix} = V\begin{pmatrix} B_2 & B_1 & A_2 \\ \iota & \iota & \iota \end{pmatrix} = -1$$

which is necessary in order for V to behave properly under odd permutation of its columns.

Table 3.2. V coefficients for three representations of D_n, just one of which is of degree 1.

E_m	E_m	A_1	A_2
c	c	$\frac{1}{\sqrt{2}}$	$\cdots$
s	s	$\frac{1}{\sqrt{2}}$	$\cdots$
c	s	$\cdots$	$\frac{1}{\sqrt{2}}$
s	c	$\cdots$	$\frac{-1}{\sqrt{2}}$

E_m	$E_{\frac{1}{2}n-m}$	B_1	B_2
c	c	$\frac{-1}{\sqrt{2}}$	$\cdots$
s	s	$\frac{1}{\sqrt{2}}$	$\cdots$
c	s	$\cdots$	$\frac{1}{\sqrt{2}}$
s	c	$\cdots$	$\frac{1}{\sqrt{2}}$

The most useful complex component system is obtained by leaving A_1, A_2, B_1, B_2 unchanged, but transforming functions belonging to an E representation according to the rule

$$\begin{aligned} g_1 &= -\frac{i}{\sqrt{2}} g_c + \frac{1}{\sqrt{2}} g_s \\ g_{-1} &= \frac{i}{\sqrt{2}} g_c + \frac{1}{\sqrt{2}} g_s \end{aligned} \tag{3.10}$$

For the complex component systems of our dihedral groups, Eq. (2.34) always takes the simple form

$$\langle a\alpha\dagger \mid g^b_{\beta\dagger} \mid c\gamma\dagger\rangle = \langle a \,\|\, g^b \,\|\, c\rangle V \begin{pmatrix} a & c & b \\ -\alpha\dagger & \gamma\dagger & \beta\dagger \end{pmatrix} \tag{3.11}$$

Tables of V in the real and complex component systems we have defined are given in Appendix D for D_4, D_5, and D_∞, and the groups D_2, D_3, and D_6 are discussed. For D_3 the coefficients appear as that part of the tables of V for the two trigonal component systems for the octahedral group O in Appendix C which refers only to the representations A_1, A_2, and E.

3.5 REDUCED MATRIX ELEMENTS

The discussion of examples of reduced matrix elements, given in Section 2.10, is modified somewhat for the D_n groups in that the vectors **l**, **s** and the set of triplet spin functions $\chi(1M_s)$ now span reducible representations.

CHAPTER 4

W Coefficients

4.1 INVARIANTS

We shall make great use of sums of products of V coefficients, referring to real component systems, which are invariant to orthogonal transformations of component systems. These sums occur in calculations because we often wish to express reduced matrix elements for coupled systems in terms of the reduced matrix elements for the constituent simpler systems. The reduced matrix elements themselves are invariant, so naturally the combinations of V coefficients occurring in equations connecting them will normally be so too.

The left-hand side of the equation

$$\sum_{\alpha\beta\gamma} V\begin{pmatrix} a & b & c \\ \alpha & \beta & \gamma \end{pmatrix}^2 = \delta(a, b, c)$$

is an example of such an invariant formed from V coefficients. We see that it is obtained by having each representation a, b, c occurring twice in the product with the same component each time. Quite apart from the fact that the sum equals $\delta(a, b, c)$, it clearly gives an invariant, for if, for example, we transform the representations to new component systems, then by Eq. (2.25) we obtain

$$\sum_{\alpha\dagger\beta\dagger\gamma\dagger} V\begin{pmatrix} a & b & c \\ \alpha\dagger & \beta\dagger & \gamma\dagger \end{pmatrix}^2$$
$$= \sum_{\substack{\alpha\beta\gamma \\ \alpha_1\beta_1\gamma_1}} A_{\alpha\dagger\alpha}B_{\beta\dagger\beta}C_{\gamma\dagger\gamma}A_{\alpha\dagger\alpha_1}B_{\beta\dagger\beta_1}C_{\gamma\dagger\gamma_1}V\begin{pmatrix} a & b & c \\ \alpha & \beta & \gamma \end{pmatrix} V\begin{pmatrix} a & b & c \\ \alpha_1 & \beta_1 & \gamma_1 \end{pmatrix}$$
$$= \sum_{\alpha\beta\gamma} V\begin{pmatrix} a & b & c \\ \alpha & \beta & \gamma \end{pmatrix}^2$$

where α_1, β_1, γ_1 are also components in the original system, and we have used the fact that $A_{\alpha\dagger\alpha}$, $B_{\beta\dagger\beta}$, and $C_{\gamma\dagger\gamma}$ are orthogonal matrices.

The foregoing example does not give a very interesting invariant, but it serves to show how we may build up more complicated ones. We take any product of V coefficients in which each representation appears twice with the same component each time. Then we sum over all components. Because each of the V coefficients contains three representations, our products must always contain an even number of factors. Clearly, there is no non-trivial invariant with only two factors. The first and most important has four factors and hence six constituent representations. We define it by the equation

$$W\begin{pmatrix} a & b & c \\ d & e & f \end{pmatrix} = \sum_{\alpha\beta\gamma\delta\epsilon\phi} V\begin{pmatrix} a & b & c \\ \alpha & \beta & \gamma \end{pmatrix} V\begin{pmatrix} a & e & f \\ \alpha & \epsilon & \phi \end{pmatrix} V\begin{pmatrix} b & f & d \\ \beta & \phi & \delta \end{pmatrix} V\begin{pmatrix} c & d & e \\ \gamma & \delta & \epsilon \end{pmatrix} \tag{4.1}$$

It might seem that this is not the only sort of invariant with four V coefficients. However, all the others either are multiples of W or turn out to be trivial invariants. For example,

$$W'\begin{pmatrix} a & b & c \\ d & e & f \end{pmatrix} = \sum_{\alpha\beta\gamma\delta\epsilon\phi} V\begin{pmatrix} a & b & c \\ \alpha & \beta & \gamma \end{pmatrix} V\begin{pmatrix} a & b & f \\ \alpha & \beta & \phi \end{pmatrix} V\begin{pmatrix} e & f & d \\ \epsilon & \phi & \delta \end{pmatrix} V\begin{pmatrix} c & d & e \\ \gamma & \delta & \epsilon \end{pmatrix}$$
$$= \lambda(c)^{-1}\delta_{cf}\,\delta(a, b, c) \sum_{\gamma\delta\epsilon} V\begin{pmatrix} e & c & d \\ \epsilon & \gamma & \delta \end{pmatrix} V\begin{pmatrix} c & d & e \\ \gamma & \delta & \epsilon \end{pmatrix}$$
$$= \lambda(c)^{-1}\delta_{cf}\,\delta(a, b, c)\delta(c, d, e)$$

where we have used Eqs. (2.18) and (2.19).

4.2 ELEMENTARY PROPERTIES OF W COEFFICIENTS

It follows at once, from the definition given by Eq. (4.1) and the corresponding property of V, that W is invariant to any even permutation of its columns. It is, however, also invariant to odd permutations, for such a one applied to the columns of W gives an odd permutation of each of the four V coefficients. So W is multiplied by

$$(-1)^{a+b+c+a+e+f+b+f+d+c+d+e} = 1$$

and is hence invariant. Finally, turning any pair of columns upside down leaves W invariant:

$$W\begin{pmatrix} a & b & c \\ d & e & f \end{pmatrix} = W\begin{pmatrix} a & e & f \\ d & b & c \end{pmatrix} = W\begin{pmatrix} d & e & c \\ a & b & f \end{pmatrix} = W\begin{pmatrix} d & b & f \\ a & e & c \end{pmatrix}$$

These relations are immediately implicit in Eq. (4.1). We have shown that there are 24 different arrangements of the six constituent representations, all of which give the same value for W. Naturally, if some of $a, \ldots, f$ are the same, then not all these W need be really different.

If one of the representations in W is A_1, then we can give a general formula for such a W.

$$\begin{aligned} W\begin{pmatrix} A_1 & b & c \\ d & e & f \end{pmatrix} &= \sum_{\beta\gamma\delta\epsilon\phi} V\begin{pmatrix} A_1 & b & c \\ \iota & \beta & \gamma \end{pmatrix} V\begin{pmatrix} A_1 & e & f \\ \iota & \epsilon & \phi \end{pmatrix} V\begin{pmatrix} b & f & d \\ \beta & \phi & \delta \end{pmatrix} V\begin{pmatrix} c & d & e \\ \gamma & \delta & \epsilon \end{pmatrix} \\ &= \lambda(b)^{-1/2}\lambda(e)^{-1/2}\delta_{bc}\delta_{ef} \sum_{\beta\delta\epsilon} V\begin{pmatrix} b & e & d \\ \beta & \epsilon & \delta \end{pmatrix} V\begin{pmatrix} b & d & e \\ \beta & \delta & \epsilon \end{pmatrix} \\ &= (-1)^{b+d+e}\lambda(b)^{-1/2}\lambda(e)^{-1/2}\delta_{bc}\delta_{ef}\,\delta(b, d, e) \end{aligned} \tag{4.2}$$

For the octahedral group there are some other special cases of interest. First, suppose that we have five T_1. Here we use Eqs. (2.23) and (2.22) to find

$$\begin{aligned} W\begin{pmatrix} T_1 & T_1 & T_1 \\ d & T_1 & T_1 \end{pmatrix} &= \frac{1}{6}\sum_{\alpha\beta\gamma\delta\epsilon\phi} \epsilon_{\alpha\beta\gamma}\epsilon_{\alpha\epsilon\phi} V\begin{pmatrix} T_1 & T_1 & d \\ \beta & \phi & \delta \end{pmatrix} V\begin{pmatrix} T_1 & d & T_1 \\ \gamma & \delta & \epsilon \end{pmatrix} \\ &= \frac{1}{6}\sum_{\beta\gamma\delta\epsilon\phi} (\delta_{\beta\epsilon}\delta_{\gamma\phi} - \delta_{\beta\phi}\delta_{\gamma\epsilon}) V\begin{pmatrix} T_1 & T_1 & d \\ \beta & \phi & \delta \end{pmatrix} V\begin{pmatrix} T_1 & d & T_1 \\ \gamma & \delta & \epsilon \end{pmatrix} \\ &= \frac{1}{6}\sum_{\beta\gamma\delta} V\begin{pmatrix} T_1 & T_1 & d \\ \beta & \gamma & \delta \end{pmatrix}^2 - \frac{1}{6}\sum_{\beta\gamma\delta} V\begin{pmatrix} T_1 & T_1 & d \\ \beta & \beta & \delta \end{pmatrix} V\begin{pmatrix} T_1 & d & T_1 \\ \gamma & \delta & \gamma \end{pmatrix} \\ &= \frac{1}{6}\delta(T_1, T_1, d) - \frac{1}{2}\delta_{dA_1} \end{aligned} \tag{4.3}$$

The same formula holds for $W\begin{pmatrix} T_1 & T_2 & T_2 \\ d & T_2 & T_2 \end{pmatrix}$, because the constituent V coefficients have identically the same values. Next, every W with six T representations in it has the value $\frac{1}{6}$. It is, perhaps, interesting to derive this value from Eq. (2.23) by general arguments. First, note that any constituent V is zero unless all the components occurring in it are different. Therefore, in Eq. (4.1), γ must be different from α and β. But so must ϕ, and hence $\gamma = \phi$ in any term that gives a non-zero contribution to W. Similarly, $\beta = \epsilon$ and $\alpha = \delta$, so

$$W = \sum_{\alpha\beta\gamma} V\begin{pmatrix} a & b & c \\ \alpha & \beta & \gamma \end{pmatrix} V\begin{pmatrix} a & e & f \\ \alpha & \beta & \gamma \end{pmatrix} V\begin{pmatrix} d & b & f \\ \alpha & \beta & \gamma \end{pmatrix} V\begin{pmatrix} d & e & c \\ \alpha & \beta & \gamma \end{pmatrix} \tag{4.4}$$

There are six terms in this sum, each of absolute value $\frac{1}{36}$, so we need to show that each is positive to achieve our result. The only possible source of a minus sign is from the V in Eqs. (2.23) which are multiples of $\epsilon_{\alpha\beta\gamma}$, i.e., those with an odd number of T_1. But 12 representations occur in each product in Eq. (4.4), so the number of V with an odd number of T_1 must be even. Hence, whether $\alpha\beta\gamma$ is an even or an odd permutation of xyz, an even number of minus signs occurs. Therefore, each term is $+\frac{1}{36}$, and $W = \frac{1}{6}$.

Next, note that W of Eq. (4.1) is necessarily zero unless

$$\delta(a, b, c)\delta(a, e, f)\delta(b, f, d)\delta(c, d, e) = 1$$

All non-zero W for the octahedral group are tabulated in Appendix C, and those for D_4, D_5, and D_6 are in Appendix D (wherein D_2 and D_3 are also discussed).

4.3 RELATION OF W TO RECOUPLING TRANSFORMATIONS

If we have three representations, say, e, f, and b, we may couple all three together. One way of doing this is first to couple e to f, thus

$$|efa\alpha\rangle = \sum_{\epsilon\phi} \langle ef\epsilon\phi \mid efa\alpha\rangle |e\epsilon\rangle |f\phi\rangle$$

giving one set of kets for each a contained in ef, and then to couple each of these to b to give

$$|ef(a), bc\gamma\rangle = \sum_{\epsilon\phi\alpha\beta} \langle ef\epsilon\phi \mid efa\alpha\rangle\langle ab\alpha\beta \mid abc\gamma\rangle |e\epsilon\rangle |f\phi\rangle |b\beta\rangle \tag{4.5}$$

We are supposing that each of the sets of kets $|e\epsilon\rangle$, $|f\phi\rangle$, and $|b\beta\rangle$ refers to quite different systems. There is no question of antisymmetrization. Further, if the representations e and f were, in fact, the same, we would introduce extra classifying parameters to distinguish the kets $|e\epsilon\rangle$ from the kets $|f\phi\rangle$. We also assume that the three constituent sets of kets are normalized. Then

$$\langle ef(a'), bc'\gamma' \mid ef(a), bc\gamma\rangle$$
$$= \sum_{\alpha\alpha'\beta\epsilon\phi} \langle ef\epsilon\phi \mid efa\alpha\rangle\langle ab\alpha\beta \mid abc\gamma\rangle\langle efa'\alpha' \mid ef\epsilon\phi\rangle\langle a'bc'\gamma' \mid a'b\alpha'\beta\rangle$$
$$= \sum_{\alpha\alpha'\beta} \delta_{aa'}\delta_{\alpha\alpha'}\langle ab\alpha\beta \mid abc\gamma\rangle\langle a'bc'\gamma' \mid a'b\alpha'\beta\rangle = \delta_{aa'}\delta_{cc'}\delta_{\gamma\gamma'}$$

where we have used the orthonormality properties for the coupling coefficients. So the coupled kets $|ef(a), bc\gamma\rangle$ are an orthonormal set.

Still keeping e, f, and b in the same order, we could have built up our coupled kets by first coupling f and b to give

$$|fbd\delta\rangle = \sum_{\phi\beta} \langle fb\phi\beta \mid fbd\delta\rangle| f\phi\rangle| b\beta\rangle$$

and then

$$|e, fb(d)c'\gamma'\rangle = \sum_{\phi\beta\epsilon\delta} \langle fb\phi\beta \mid fbd\delta\rangle\langle ed\epsilon\delta \mid edc'\gamma'\rangle| e\epsilon\rangle| f\phi\rangle| b\beta\rangle \tag{4.6}$$

For rather obvious reasons, the elements of the matrix of transformation between the kets of Eqs. (4.5) and (4.6) are called recoupling matrix elements. Because of the group orthogonality relations for integrals, they are zero unless $c = c'$ and $\gamma = \gamma'$. When the latter equalities hold, the elements are independent of γ. Their only dependence on γ and γ' is through a factor $\delta_{\gamma\gamma'}$.

The recoupling matrix elements are multiples of W coefficients, as we now see. In fact,

$$\delta_{cc'}\delta_{\gamma\gamma'}\langle e, fb(d)c\gamma \mid ef(a), bc\gamma\rangle = \langle e, fb(d)c'\gamma' \mid ef(a), bc\gamma\rangle$$
$$= \sum_{\alpha\beta\delta\epsilon\phi} \langle fb\phi\beta \mid fbd\delta\rangle\langle ed\epsilon\delta \mid edc'\gamma'\rangle\langle ef\epsilon\phi \mid efa\alpha\rangle\langle ab\alpha\beta \mid abc\gamma\rangle \tag{4.7}$$
$$= \lambda(c)^{1/2}\lambda(c')^{1/2}\lambda(a)^{1/2}\lambda(d)^{1/2}$$
$$\sum_{\alpha\beta\delta\epsilon\phi} V\begin{pmatrix} a & b & c \\ \alpha & \beta & \gamma \end{pmatrix} V\begin{pmatrix} a & e & f \\ \alpha & \epsilon & \phi \end{pmatrix} V\begin{pmatrix} b & d & f \\ \beta & \delta & \phi \end{pmatrix} V\begin{pmatrix} c' & e & d \\ \gamma' & \epsilon & \delta \end{pmatrix}$$

Putting $c = c'$, $\gamma = \gamma'$ in Eq. (4.7), rearranging the columns of the last two V coefficients, and summing over γ gives

$$\langle e, fb(d)c\gamma \mid ef(a), bc\gamma\rangle = (-1)^{b+c+e+f}\lambda(a)^{1/2}\lambda(d)^{1/2}W\begin{pmatrix} a & b & c \\ d & e & f \end{pmatrix} \tag{4.8}$$

Introducing Eq. (4.8) into Eq. (4.7), we have

$$\sum_{\alpha\beta\delta\epsilon\phi} V\begin{pmatrix} a & b & c \\ \alpha & \beta & \gamma \end{pmatrix} V\begin{pmatrix} a & e & f \\ \alpha & \epsilon & \phi \end{pmatrix} V\begin{pmatrix} b & f & d \\ \beta & \phi & \delta \end{pmatrix} V\begin{pmatrix} c' & d & e \\ \gamma' & \delta & \epsilon \end{pmatrix}$$
$$= \lambda(c)^{-1}\delta_{cc'}\delta_{\gamma\gamma'}W\begin{pmatrix} a & b & c \\ d & e & f \end{pmatrix} \tag{4.9}$$

This all-important equation appears not to follow from the definition of W and the elementary properties of the V; we considered the recoupling matrices in order to obtain it. Now that we possess it, we no longer need to use the recoupling matrices. However, we shall meet them again when we come to discuss spin-orbit coupling in n-electron systems.

4.4 EQUATIONS SATISFIED BY W COEFFICIENTS

Equation (4.9) is but one of a series of identities involving V and W coefficients. It may be called a (4, 0) equation because it has four V coefficients on the left, but none on the right to multiply the W. We now deduce a (3, 1) equation by multiplying Eq. (4.9) through by

$$\lambda(c)V\begin{pmatrix} a & b & c \\ \alpha' & \beta' & \gamma \end{pmatrix}$$

and summing over c and γ. The left-hand side is then simplified by using the second of Eqs. (2.18). We drop the prime signs on α', β', c', γ' in the resulting equation to obtain finally

$$\sum_{\delta\epsilon\phi} V\begin{pmatrix} a & e & f \\ \alpha & \epsilon & \phi \end{pmatrix} V\begin{pmatrix} b & f & d \\ \beta & \phi & \delta \end{pmatrix} V\begin{pmatrix} c & d & e \\ \gamma & \delta & \epsilon \end{pmatrix} = W\begin{pmatrix} a & b & c \\ d & e & f \end{pmatrix} V\begin{pmatrix} a & b & c \\ \alpha & \beta & \gamma \end{pmatrix} \tag{4.10}$$

The (2, 2) equation is now derived from Eq. (4.10) in exactly the same way by multiplying through by

$$\lambda(c)V\begin{pmatrix} c & d & e \\ \gamma & \delta' & \epsilon' \end{pmatrix}$$

and summing over c and γ. We obtain

$$\sum_{\phi} V\begin{pmatrix} a & e & f \\ \alpha & \epsilon & \phi \end{pmatrix} V\begin{pmatrix} b & f & d \\ \beta & \phi & \delta \end{pmatrix} = \sum_{c\gamma} \lambda(c) W\begin{pmatrix} a & b & c \\ d & e & f \end{pmatrix} V\begin{pmatrix} a & b & c \\ \alpha & \beta & \gamma \end{pmatrix} V\begin{pmatrix} c & d & e \\ \gamma & \delta & \epsilon \end{pmatrix} \tag{4.11}$$

We obtain the (1, 3) equation by multiplying Eq. (4.11) by

$$V\begin{pmatrix} a & e & g \\ \alpha & \epsilon & \eta \end{pmatrix}$$

and summing over α and ϵ, using the first of Eqs. (2.18). The left-hand side becomes

$$\sum_{\phi\alpha\epsilon} V\begin{pmatrix} a & e & g \\ \alpha & \epsilon & \eta \end{pmatrix} V\begin{pmatrix} a & e & f \\ \alpha & \epsilon & \phi \end{pmatrix} V\begin{pmatrix} b & f & d \\ \beta & \phi & \delta \end{pmatrix}$$

$$= \sum_{\phi} \lambda(f)^{-1}\delta_{fg}\delta_{\phi\eta}\delta(a, e, g) V\begin{pmatrix} b & f & d \\ \beta & \phi & \delta \end{pmatrix} \tag{4.12}$$

$$= \lambda(f)^{-1}\delta_{fg}\delta(a, e, g) V\begin{pmatrix} b & g & d \\ \beta & \eta & \delta \end{pmatrix}$$

$$= \sum_{c\gamma\alpha\epsilon} \lambda(c) W\begin{pmatrix} a & b & c \\ d & e & f \end{pmatrix} V\begin{pmatrix} a & b & c \\ \alpha & \beta & \gamma \end{pmatrix} V\begin{pmatrix} c & d & e \\ \gamma & \delta & \epsilon \end{pmatrix} V\begin{pmatrix} a & e & g \\ \alpha & \epsilon & \eta \end{pmatrix}$$

where the last line is the right-hand side.

The (0, 4) equation is one of three useful equations satisfied by W alone. We multiply Eq. (4.12) through by

$$V\begin{pmatrix} b & g & d \\ \beta & \eta & \delta \end{pmatrix}$$

and sum over β, η, and δ to give

$$\sum_{c} \lambda(c) W\begin{pmatrix} a & b & c \\ d & e & f \end{pmatrix} W\begin{pmatrix} a & b & c \\ d & e & g \end{pmatrix} = \lambda(f)^{-1}\delta_{fg}\delta(a, e, g)\delta(b, d, g) \tag{4.13}$$

As well as this orthogonality rule, there is a somewhat similar one called the associative law. The latter is established by introducing

$$\sum_{\alpha\beta\gamma} V\begin{pmatrix} a & b & c \\ \alpha & \beta & \gamma \end{pmatrix}^2 = 1$$

into the left-hand side of the equation we want to prove and then using Eq. (4.10) as follows:

$$\sum_{c} (-1)^c \lambda(c) W\begin{pmatrix} a & b & c \\ d & e & f \end{pmatrix} W\begin{pmatrix} a & b & c \\ e & d & g \end{pmatrix}$$

$$= \sum_{c\alpha\beta\gamma} (-1)^c \lambda(c) W\begin{pmatrix} a & b & c \\ d & e & f \end{pmatrix} V\begin{pmatrix} a & b & c \\ \alpha & \beta & \gamma \end{pmatrix} W\begin{pmatrix} a & b & c \\ e & d & g \end{pmatrix} V\begin{pmatrix} a & b & c \\ \alpha & \beta & \gamma \end{pmatrix}$$

$$= \sum_{\substack{c\alpha\beta\gamma \\ \delta\epsilon\phi\delta'\epsilon'\eta}} (-1)^c \lambda(c) V\begin{pmatrix} a & e & f \\ \alpha & \epsilon & \phi \end{pmatrix} V\begin{pmatrix} b & f & d \\ \beta & \phi & \delta \end{pmatrix} V\begin{pmatrix} c & d & e \\ \gamma & \delta & \epsilon \end{pmatrix} V\begin{pmatrix} a & d & g \\ \alpha & \delta' & \eta \end{pmatrix} \tag{4.14}$$

$$V\begin{pmatrix} b & g & e \\ \beta & \eta & \epsilon' \end{pmatrix} V\begin{pmatrix} c & e & d \\ \gamma & \epsilon' & \delta' \end{pmatrix}$$

$$= \sum_{\alpha\beta\delta\epsilon\phi\eta} (-1)^{d+e} V\begin{pmatrix} a & e & f \\ \alpha & \epsilon & \phi \end{pmatrix} V\begin{pmatrix} b & f & d \\ \beta & \phi & \delta \end{pmatrix} V\begin{pmatrix} a & d & g \\ \alpha & \delta & \eta \end{pmatrix} V\begin{pmatrix} b & g & e \\ \beta & \eta & \epsilon \end{pmatrix}$$

$$= (-1)^{f+g} W\begin{pmatrix} a & d & g \\ b & e & f \end{pmatrix}$$

If we put $g = A_1$ in Eq. (4.14) and use Eq. (4.2), we readily obtain

$$\sum_c \lambda(c) W \begin{pmatrix} a & b & c \\ a & b & f \end{pmatrix} = \delta(a, b, c) \tag{4.15}$$

which is sometimes useful.

The final equation was found in the theory for spherical symmetry by Biedenharn and Elliott and relates a product of two W to a sum over three W. Here we start off as for the associative law, obtaining

$$W \begin{pmatrix} a & b & c \\ d & e & f \end{pmatrix} W \begin{pmatrix} a & b & c \\ \bar{d} & \bar{e} & \bar{f} \end{pmatrix}$$

$$= \sum_{\alpha\beta\gamma\delta\epsilon\phi\bar{\delta}\bar{\epsilon}\bar{\phi}} V \begin{pmatrix} a & e & f \\ \alpha & \epsilon & \phi \end{pmatrix} V \begin{pmatrix} b & f & d \\ \beta & \phi & \delta \end{pmatrix} V \begin{pmatrix} c & d & e \\ \gamma & \delta & \epsilon \end{pmatrix} V \begin{pmatrix} a & \bar{e} & \bar{f} \\ \alpha & \bar{\epsilon} & \bar{\phi} \end{pmatrix}$$

$$V \begin{pmatrix} b & \bar{f} & \bar{d} \\ \beta & \bar{\phi} & \bar{\delta} \end{pmatrix} V \begin{pmatrix} c & \bar{d} & \bar{e} \\ \gamma & \bar{\delta} & \bar{\epsilon} \end{pmatrix}$$

$$= (-1)^{c+d+e} \sum_{g\alpha\beta\eta\delta\epsilon\phi\bar{\delta}\bar{\epsilon}\bar{\phi}} \lambda(g) V \begin{pmatrix} a & e & f \\ \alpha & \epsilon & \phi \end{pmatrix} V \begin{pmatrix} a & \bar{e} & \bar{f} \\ \alpha & \bar{\epsilon} & \bar{\phi} \end{pmatrix} V \begin{pmatrix} g & \bar{e} & e \\ \eta & \bar{\epsilon} & \epsilon \end{pmatrix} W \begin{pmatrix} d & \bar{d} & g \\ \bar{e} & e & c \end{pmatrix}$$

$$V \begin{pmatrix} d & \bar{d} & g \\ \delta & \bar{\delta} & \eta \end{pmatrix} V \begin{pmatrix} b & f & d \\ \beta & \phi & \delta \end{pmatrix} V \begin{pmatrix} b & \bar{f} & \bar{d} \\ \beta & \bar{\phi} & \bar{\delta} \end{pmatrix}$$

where we have used Eq. (4.11) and rearranged somewhat. Each product of three V coefficients is now simplified by using Eq. (4.10). Finally, Eq. (2.19) eliminates the V altogether, and we have

$$W \begin{pmatrix} a & b & c \\ d & e & f \end{pmatrix} W \begin{pmatrix} a & b & c \\ \bar{d} & \bar{e} & \bar{f} \end{pmatrix} = (-1)^{a+b+c+d+e+f+\bar{d}+\bar{e}+\bar{f}}$$

$$\sum_g (-1)^g \lambda(g) W \begin{pmatrix} g & e & \bar{e} \\ a & \bar{f} & f \end{pmatrix} W \begin{pmatrix} g & f & \bar{f} \\ b & \bar{d} & d \end{pmatrix} W \begin{pmatrix} g & d & \bar{d} \\ c & \bar{e} & e \end{pmatrix} \tag{4.16}$$

CHAPTER 5

Irreducible Products and Their Matrix Elements

5.1 DEFINITION OF AN IRREDUCIBLE PRODUCT

Just as pairs of sets of kets or bras may be coupled together to form bases for irreducible representations contained in a direct product, so also may operators. Suppose A^a_α and B^b_β are operator sets spanning the irreducible representations a and b respectively. Then the quantities

$$(A^a \times B^b)^c_\gamma = \sum_{\alpha\beta} \langle ab\alpha\beta \mid abc\gamma\rangle A^a_\alpha B^b_\beta = \lambda(c)^{1/2} \sum_{\alpha\beta} V\begin{pmatrix} a & b & c \\ \alpha & \beta & \gamma \end{pmatrix} A^a_\alpha B^b_\beta \tag{5.1}$$

span the irreducible representation c, if c is contained in ab, and are zero otherwise. The expression on the left of Eq. (5.1) is called an irreducible product; we note that there is no ambiguity about the notation, because c can never be contained in ab more than once.

The behaviour of an irreducible product when we interchange A and B follows from the properties of the V coefficients. Should each component of A^a_α commute with each component of B^b_β, we find

$$(A^a \times B^b)^c_\gamma = (-1)^{a+b+c}(B^b \times A^a)^c_\gamma \tag{5.2}$$

When $A^a_\alpha = B^b_\beta$ and the components of A^a_α mutually commute, it follows from Eq. (5.2) that

$$(A^a \times A^a)^c_\gamma = 0$$

when c is in the antisymmetrized square of a. For the octahedral group O, if A^a_α is an angular momentum ($a = T_1$), then the only such c is T_1, and in this case

$$(A^a \times A^a)^{T_1}_\gamma = -\frac{1}{\sqrt{2}}\sum_{\alpha\beta} \epsilon_{\alpha\beta\gamma}A^a_\alpha A^a_\beta = -\frac{1}{\sqrt{2}}(\mathbf{A}^a \wedge \mathbf{A}^a)_\gamma = -\frac{i}{\sqrt{2}}A^a_\gamma \tag{5.3}$$

Irreducible products are of wide occurrence, and all the most important operators whose matrix elements we wish to evaluate can be written naturally as sums of irreducible products of some sort or other. Many of these are expressible in terms of one or the other of two special cases of Eq. (5.1). The first special case occurs when $c = A_1$. It is convenient to introduce here the notation

$$\mathbf{A}^a \cdot \mathbf{B}^a = \sum_\alpha A^a_\alpha B^a_\alpha \tag{5.4}$$

which is an obvious generalization of the scalar product for vectors and is actually equal to it when $\mathbf{A}$ and $\mathbf{B}$ are vectors, and a is the T_1 representation of the octahedral group. Then from Eqs. (2.20) and (5.1) we find

$$(A^a \times B^b)^{A_1} = \delta_{ab}\lambda(a)^{-1/2}\mathbf{A}^a \cdot \mathbf{B}^a \tag{5.5}$$

where we have dropped the unnecessary suffix ι on the left-hand side. The piece $\mathbf{s}(i) \cdot \mathbf{u}(i)$ of the spin-orbit coupling energy for the group O [often $\mathbf{u} = \xi(r)\mathbf{l}$] belonging to the ith electron is a good example here; we see that

$$\mathbf{s}(i) \cdot \mathbf{u}(i) = \sqrt{3}\,(s(i) \times u(i))^{A_1} \tag{5.6}$$

where $a = b = T_1$ is understood.

The second special case occurs when either $b = A_1$ and $a = c$ or $a = A_1$ and $b = c$. Very often the operator having symmetry A_1 will then be just the number 1. It may seem at the moment that to introduce the concept of irreducible product in such a case is merely a pointless complication. However, as later chapters will show, this concept enables us to fit all our calculations into a single unified scheme and does not, in fact, create any extra difficulty. As with the spin-orbit coupling, so we often wish to factorize other operators into first parts dealing with spin functions alone and second parts dealing with space functions alone, even when they refer explicitly only to spin or only to space. For example, for O we shall write

$$\begin{aligned} l_\beta &= (1 \times l)_\beta^{T_1} \\ s_\alpha &= (s \times 1)_\alpha^{T_1} \end{aligned} \tag{5.7}$$

for orbital and spin angular momenta in accordance with Eq. (5.1).

Further examples are easy to find. However, first it is helpful to examine some properties of irreducible products of three sets of operators. We build these up in just the same way as we do for sets of kets or bras; again, a question of order of coupling arises. In other words, we have both

$$[(A^a \times B^b)^d \times C^c]^f_\phi = \lambda(f)^{1/2}\lambda(d)^{1/2} \sum_{\alpha\beta\gamma\delta} V\begin{pmatrix} d & c & f \\ \delta & \gamma & \phi \end{pmatrix} V\begin{pmatrix} a & b & d \\ \alpha & \beta & \delta \end{pmatrix} A^a_\alpha B^b_\beta C^c_\gamma \tag{5.8}$$

and

$$[A^a \times (B^b \times C^c)^e]^f_\phi = \lambda(f)^{1/2}\lambda(e)^{1/2} \sum_{\alpha\beta\gamma\epsilon} V\begin{pmatrix} b & c & e \\ \beta & \gamma & \epsilon \end{pmatrix} V\begin{pmatrix} a & e & f \\ \alpha & \epsilon & \phi \end{pmatrix} A^a_\alpha B^b_\beta C^c_\gamma \tag{5.9}$$

connected by the pair of equations

$$\begin{aligned} [A^a \times (B^b \times C^c)^e]^f_\phi &= (-1)^{a+b+c+f}\lambda(e)^{1/2} \sum_d \lambda(d)^{1/2} W\begin{pmatrix} b & a & d \\ f & c & e \end{pmatrix} \\ &\qquad [(A^a \times B^b)^d \times C^c]^f_\phi \\ [(A^a \times B^b)^d \times C^c]^f_\phi &= (-1)^{a+b+c+f}\lambda(d)^{1/2} \sum_e \lambda(e)^{1/2} W\begin{pmatrix} b & a & d \\ f & c & e \end{pmatrix} \\ &\qquad [A^a \times (B^b \times C^c)^e]^f_\phi \end{aligned} \tag{5.10}$$

Equations (5.8) and (5.9) follow directly from expansion. Then the first of Eqs. (5.10) is derived by using Eq. (4.11) in the form

$$\sum_\epsilon V\begin{pmatrix} b & c & e \\ \beta & \gamma & \epsilon \end{pmatrix} V\begin{pmatrix} a & e & f \\ \alpha & \epsilon & \phi \end{pmatrix} = \sum_{d\delta} \lambda(d) W\begin{pmatrix} b & a & d \\ f & c & e \end{pmatrix} V\begin{pmatrix} b & a & d \\ \beta & \alpha & \delta \end{pmatrix} V\begin{pmatrix} d & f & c \\ \delta & \phi & \gamma \end{pmatrix}$$

The second of (5.10) is derived from the first by Eq. (4.13). When $f = A_1$, we use Eq. (4.2) to see that the equations simplify to give the formula

$$[A^a \times (B^b \times C^c)^a]^{A_1} = [(A^a \times B^b)^c \times C^c]^{A_1} \tag{5.11}$$

5.2 VARIOUS SPECIAL OPERATORS

In using the theoretical method we are developing, it is very important to be able to introduce easily into the scheme the operators whose matrix elements one wishes to calculate. This means expressing them as irreducible products, each constituent element of which refers to an independent part of the system. The choice of suitable parts is largely determined by the type of coupling scheme adopted for the kets and bras. Thus, for example, if the kets and bras are each factorizable into a spin part and a space part (as in Eq. (2.41)) we also write our operators as sums of irreducible products of spin operators and spatial operators, as in Eqs. (5.6) and (5.7).

As another illustration,* we now reproduce here the results of section 12.2 of Griffith (26) to show in outline how various small refinements to the Hamiltonian may be got into the right form, using the group O as an illustration. In (26), quantities $V_k(\mathbf{T}_1, \mathbf{T}_2)$ are introduced. Actually, they are components of an irreducible product of two vectors referring to the three-dimensional rotation group, although we shall not pursue that matter here. Then in Eq. (12.5) of (26) a quantity is defined which we will write here as

$$Q = \sum_{k=1}^{5} V_k(\mathbf{A}, \mathbf{A}) V_k(\mathbf{B}, \mathbf{C}) \tag{5.12}$$

The nuclear hyperfine and quadrupole interactions and the spin-spin coupling are all expressed in terms of suitable Q in such a way that $\mathbf{A}$ is always a spatial operator and $\mathbf{B}$ and $\mathbf{C}$ are each either electronic or nuclear spin vectors.

The incorporation of Q in our scheme is now easy once we notice the relations

$$\begin{aligned}
(A \times B)^{E}_{\theta} &= -\frac{1}{\sqrt{2}} V_1(\mathbf{A}, \mathbf{B}) \\
(A \times B)^{E}_{\epsilon} &= -\frac{1}{\sqrt{2}} V_2(\mathbf{A}, \mathbf{B}) \\
(A \times B)^{T_2}_{x} &= -\frac{1}{\sqrt{2}} V_4(\mathbf{A}, \mathbf{B}) \\
(A \times B)^{T_2}_{y} &= -\frac{1}{\sqrt{2}} V_5(\mathbf{A}, \mathbf{B}) \\
(A \times B)^{T_2}_{z} &= -\frac{1}{\sqrt{2}} V_3(\mathbf{A}, \mathbf{B})
\end{aligned} \tag{5.13}$$

* The rest of the present section may be omitted without disturbing the continuity of the book.

for any pair of vectors **A** and **B**. Then

$$Q = 2 \sum_{\gamma;c=E,T_2} (A \times A)^c_\gamma (B \times C)^c_\gamma$$
$$= 2\lambda(c)^{1/2} \sum_{c=E,T_2} [(A \times A)^c \times (B \times C)^c]^{A_1} \tag{5.14}$$

which gives already the correct form for the nuclear quadrupole interaction and the spin-spin coupling, if we use Eqs. (12.27) and (12.30) of ref. (26). For the nuclear hyperfine interaction, we must evaluate over the electronic states the matrix of the coefficient of each component of the nuclear spin **I** in the expression [(26), Eq. (12.16)].

$$b = -\frac{3}{2}\xi\overline{r^{-3}} \sum_{k=1}^{5} V_k(\mathbf{l}, \mathbf{l})V_k(\mathbf{s}, \mathbf{I})$$

However, using Eqs. (5.14) and (5.11), we have

$$b = -3\xi\overline{r^{-3}} \sum_{c=E,T_2} [(l \times l)^c \times (s \times I)^c]^{A_1}$$
$$= -3\xi\overline{r^{-3}} \sum_{c=E,T_2} [(l \times l)^c \times s)^{T_1} \times I]^{A_1} \tag{5.15}$$

which achieves our object. Note that here there are two reduced matrix elements for $l \times l$, one each for $c = E$ and $c = T_2$, and that these are independent as far as operations entirely within the octahedral group are concerned.

An interesting little relation is

$$\sum_{g\eta} (A^a \times B^a)^g_\eta (C^a \times D^a)^g_\eta = (\mathbf{A}^a \cdot \mathbf{C}^a)(\mathbf{B}^a \cdot \mathbf{D}^a) \tag{5.16}$$

providing B^a_β commutes with C^a_γ. Equation (5.16) follows immediately on expansion using Eq. (2.18). If we put $A^a_\alpha = B^a_\alpha$ and assume that components of A^a_α all mutually commute, then the sum in Eq. (5.16) may be restricted to those g contained in the symmetrized square of a. Now we put $a = T_1$, and we find that Q of Eq. (5.14) lacks only the term $c = A_1$ to be a multiple of a sum of the kind occurring in Eq. (5.16). Hence

$$\tfrac{1}{2}Q + \tfrac{1}{3}A^2(\mathbf{B} \cdot \mathbf{C}) = (\mathbf{A} \cdot \mathbf{B})(\mathbf{A} \cdot \mathbf{C})$$

or

$$Q = -\tfrac{2}{3}\mathbf{A}^2(\mathbf{B} \cdot \mathbf{C}) + 2(\mathbf{A} \cdot \mathbf{B})(\mathbf{A} \cdot \mathbf{C}) \tag{5.17}$$

which is Eq. (12.7) of ref. (26) in a different notation. If, however, **A** is an angular momentum, then we must include the term $c = T_1$ and add to the right-hand side of Eq. (5.17):

$$-2 \sum_{\gamma} (A \times A)^{T_1}_\gamma (B \times C)^{T_1}_\gamma = -i\mathbf{A} \cdot \mathbf{B} \wedge \mathbf{C}$$

as discussed in (26), p. 324, by a clumsier method.

Finally, note that the electrostatic interaction can also be expanded in terms of irreducible products. This was done in effect in Griffith (26) Eq. (9.45) [see also Jarrett (30), Parr (32)], where it was written in the form

$$\mathcal{V} = \sum_{ia\alpha} \sum_{\kappa<\lambda} g_{ia\alpha}(\kappa) g_{ia\alpha}(\lambda)$$

where κ, λ number the electrons, a are irreducible representations and α number the components. In our new notation, this equation would read

$$\mathcal{V} = \sum_{ia} \sum_{\kappa<\lambda} \lambda(a)^{1/2}[g_{ia}(\kappa) \times g_{ia}(\lambda)]^{A_1}$$

Certain parts of Sections 9.7.5–9.7.7 in Griffith (26) can be expressed more concisely in our new notation. For example, Eq. (9.48) becomes simply

$$\langle e\alpha \mid g_\beta \mid e\alpha' \rangle = 2pV \begin{pmatrix} E & E & E \\ \alpha & \alpha' & \beta \end{pmatrix}$$

However, this method has not been much used yet to treat the electrostatic interaction, although it is potentially capable of it [Tanabe (22), Parr (32)].

5.3 MATRIX ELEMENTS

We now suppose that we have a system consisting of coupled parts, e.g.,

$$|abc\gamma\rangle = \sum_{\alpha\beta} \langle ab\alpha\beta \mid abc\gamma\rangle | ab\alpha\beta\rangle \tag{5.18}$$

where the kets $|ab\alpha\beta\rangle$ are simple products of kets $|a\alpha\rangle$, $|b\beta\rangle$ referring to independent systems or parts of a system. An important example here occurs for two-electron systems [Eq. (2.40)], where we take $a\alpha$ to refer to spin coordinates and $b\beta$ to space coordinates. Then an operator D^d_δ referring only to the first part of the system has matrix elements satisfying

$$\langle ab\alpha\beta \mid D^d_\delta \mid a'b'\alpha'\beta'\rangle = \langle a\alpha \mid D^d_\delta \mid a'\alpha'\rangle\langle b\beta \mid b'\beta'\rangle \tag{5.19}$$

Similarly, if E^e_ϵ refers to the second part,

$$\langle ab\alpha\beta \mid E^e_\epsilon \mid a'b'\alpha'\beta'\rangle = \langle a\alpha \mid a'\alpha' \rangle \langle b\beta \mid E^e_\epsilon \mid b'\beta'\rangle \tag{5.20}$$

and, finally,

$$\begin{aligned} \langle ab\alpha\beta \mid D^d_\delta E^e_\epsilon \mid a'b'\alpha'\beta'\rangle &= \langle a\alpha \mid D^d_\delta \mid a'\alpha'\rangle\langle b\beta \mid E^e_\epsilon \mid b'\beta'\rangle \\ &= \langle a \| D^d \| a'\rangle\langle b \| E^e \| b'\rangle V \begin{pmatrix} a & a' & d \\ \alpha & \alpha' & \delta \end{pmatrix} V \begin{pmatrix} b & b' & e \\ \beta & \beta' & \epsilon \end{pmatrix} \end{aligned} \tag{5.21}$$

An irreducible product of D^d_δ and E^e_ϵ will have reduced matrix elements within the system of coupled states according to the formula

$$\langle abc\gamma \mid (D^d \times E^e)^f_\phi \mid a'b'c'\gamma'\rangle$$
$$= \langle abc \,\|\, (D^d \times E^e)^f \,\|\, a'b'c'\rangle V\begin{pmatrix} c & c' & f \\ \gamma & \gamma' & \phi \end{pmatrix} \quad (5.22)$$

The matrix element on the left-hand side can be expanded by using Eqs. (5.1) and (5.18); it can then be expressed in terms of the reduced matrix elements of D^d and E^e within the two separate parts of the system by Eq. (5.21). Thus, after inverting Eq. (5.22) by using Eq. (2.19), we find

$$\langle abc \,\|\, (D^d \times E^e)^f \,\|\, a'b'c'\rangle = \sum_{\gamma\phi\gamma'} \langle abc\gamma \mid (D^d \times E^e)^f_\phi \mid a'b'c'\gamma'\rangle V\begin{pmatrix} c & c' & f \\ \gamma & \gamma' & \phi \end{pmatrix}$$
$$= \sum_{\alpha\beta\gamma\alpha'\beta'\gamma'\delta\epsilon\phi} \langle abc\gamma \mid ab\alpha\beta\rangle\langle de\delta\epsilon \mid def\phi\rangle\langle a'b'\alpha'\beta' \mid a'b'c'\gamma'\rangle$$
$$V\begin{pmatrix} c & c' & f \\ \gamma & \gamma' & \phi \end{pmatrix} \langle ab\alpha\beta \mid D^d_\delta E^e_\epsilon \mid a'b'\alpha'\beta'\rangle \quad (5.23)$$
$$= \lambda(c)^{1/2}\lambda(f)^{1/2}\lambda(c')^{1/2}\langle a \,\|\, D^d \,\|\, a'\rangle\langle b \,\|\, E^e \,\|\, b'\rangle \sum_{\alpha\beta\gamma\alpha'\beta'\gamma'\delta\epsilon\phi} V\begin{pmatrix} a & b & c \\ \alpha & \beta & \gamma \end{pmatrix}$$
$$V\begin{pmatrix} d & e & f \\ \delta & \epsilon & \phi \end{pmatrix} V\begin{pmatrix} a' & b' & c' \\ \alpha' & \beta' & \gamma' \end{pmatrix} V\begin{pmatrix} c & c' & f \\ \gamma & \gamma' & \phi \end{pmatrix} V\begin{pmatrix} a & a' & d \\ \alpha & \alpha' & \delta \end{pmatrix} V\begin{pmatrix} b & b' & e \\ \beta & \beta' & \epsilon \end{pmatrix}$$

The sum in Eq. (5.23) is an invariant constructed from six V coefficients, and it is written concisely as

$$X\begin{bmatrix} a & b & c \\ a' & b' & c' \\ d & e & f \end{bmatrix}$$

Its properties are examined in Chapter 8. If $f = A_1$, $d = e$, and $c = c'$, the sum reduces to only four terms, becoming

$$\lambda(c)^{-1/2}\lambda(d)^{-1/2} \sum_{\alpha\beta\gamma\alpha'\beta'\delta} V\begin{pmatrix} a & b & c \\ \alpha & \beta & \gamma \end{pmatrix} V\begin{pmatrix} a' & b' & c \\ \alpha' & \beta' & \gamma \end{pmatrix} V\begin{pmatrix} a & a' & d \\ \alpha & \alpha' & \delta \end{pmatrix} V\begin{pmatrix} b & b' & d \\ \beta & \beta' & \delta \end{pmatrix}$$
$$= (-1)^{a'+b+c+d}\lambda(c)^{-1/2}\lambda(d)^{-1/2} W\begin{pmatrix} a' & a & d \\ b & b' & c \end{pmatrix}$$

so, in terms of the scalar product,

$$\langle abc\gamma \mid \mathbf{D}^d \cdot \mathbf{E}^d \mid a'b'c\gamma\rangle$$
$$= (-1)^{a'+b+c+d}\langle a \,\|\, D^d \,\|\, a'\rangle\langle b \,\|\, E^d \,\|\, b'\rangle W\begin{pmatrix} a & b & c \\ b' & a' & d \end{pmatrix} \quad (5.24)$$

If any representation symbol in Eq. (5.23) is A_1, the sum reduces to a multiple of a W coefficient. Two important cases occur when $D^d = 1$ or $E^e = 1$ so that we are really evaluating just the matrix of an operator belonging to a single part of the system. The formula given by Eq. (2.35) for the reduced matrix element of a pure number enables us to eliminate one of the reduced matrix elements in Eq. (5.23) completely, and we obtain

$$\langle abc \| D^d \| a'b'c' \rangle$$
$$= (-1)^{a+b+c'+d}\delta_{bb'}\lambda(c)^{1/2}\lambda(c')^{1/2}\langle a \| D^d \| a' \rangle W \begin{pmatrix} a' & a & d \\ c & c' & b \end{pmatrix} \quad (5.25)$$

$$\langle abc \| E^e \| a'b'c' \rangle$$
$$= (-1)^{a+b'+c+e}\delta_{aa'}\lambda(c)^{1/2}\lambda(c')^{1/2}\langle b \| E^e \| b' \rangle W \begin{pmatrix} b' & b & e \\ c & c' & a \end{pmatrix} \quad (5.26)$$

5.4 SECOND-ORDER PERTURBATION THEORY AND AN OPERATOR EQUIVALENT

Suppose now that the Hamiltonian $\mathcal{H}$ for our set of coupled states is the sum

$$\mathcal{H} = \mathcal{H}_0 + \mathcal{H}_1$$

of an "unperturbed" part $\mathcal{H}_0$ which is entirely diagonal within them, i.e., satisfies

$$\langle ab\alpha\beta \mid \mathcal{H}_0 \mid a'b'\alpha'\beta' \rangle = \langle ab \| \mathcal{H}_0 \| ab \rangle \lambda(a)^{-1/2}\lambda(b)^{-1/2}\delta_{aa'}\delta_{bb'}\delta_{\alpha\alpha'}\delta_{\beta\beta'}$$

and $\mathcal{H}_1$, which is small and of the form

$$\mathcal{H}_1 = \mathbf{D}^d \cdot \mathbf{E}^d \quad (5.27)$$

So $\mathcal{H}$ commutes with all elements of O, and in finding the modification to a state $|abc\gamma\rangle$ we need consider only states like $|a'b'c\gamma\rangle$.

The energy of $|abc\gamma\rangle$ correct to second order is

$$E = \langle abc\gamma \mid \mathcal{H} \mid abc\gamma \rangle - \sum_n E_n^{-1} \mid \langle abc\gamma \mid \mathcal{H}_1 \mid na'b'c\gamma \rangle|^2 \quad (5.28)$$

where n classifies the excited states and E_n is the energy of the nth above the state $|abc\gamma\rangle$. When we use Eq. (5.24), the first-order term is

$$E_1 = (-1)^{a+b+c+d}\langle a \| D^d \| a \rangle \langle b \| E^d \| b \rangle W \begin{pmatrix} a & a & d \\ b & b & c \end{pmatrix} \quad (5.29)$$

and the second-order term is

$$E_2 = -\sum_n E_n^{-1} |\langle a \| D^d \| na' \rangle|^2 \, |\langle b \| E^d \| nb' \rangle|^2 W \begin{pmatrix} a & b & c \\ b' & a' & d \end{pmatrix}^2$$
$$= -\sum_n Q_n W \begin{pmatrix} a & b & c \\ b' & a' & d \end{pmatrix}^2 \quad (5.30)$$

say. It is interesting now to use the Biedenharn-Elliott identity [Eq. (4.16)] with $d = \bar{d}$, $e = \bar{e}$, $f = \bar{f}$ to give us

$$W\begin{pmatrix} a & b & c \\ b' & a' & d \end{pmatrix}^2 = W\begin{pmatrix} b' & a' & c \\ a & b & d \end{pmatrix}^2$$
$$= (-1)^{a'+b'+c} \sum_g (-1)^g \lambda(g) W\begin{pmatrix} g & b & b \\ b' & d & d \end{pmatrix} W\begin{pmatrix} g & d & d \\ a' & a & a \end{pmatrix} W\begin{pmatrix} a & a & g \\ b & b & c \end{pmatrix}$$

and

$$E_2 = -\sum_n (-1)^{a'+b'+c} Q_n \sum_g (-1)^g \lambda(g) W\begin{pmatrix} g & b & b \\ b' & d & d \end{pmatrix} W\begin{pmatrix} g & d & d \\ a' & a & a \end{pmatrix} W\begin{pmatrix} a & a & g \\ b & b & c \end{pmatrix} \tag{5.31}$$

What have we done? By obtaining an expression linear in

$$(-1)^c W\begin{pmatrix} a & a & g \\ b & b & c \end{pmatrix}$$

with coefficients independent of c, we have shown that E_2 mimics the first-order effects of a set of scalar products, one product for each g occurring in the sum. Equation (5.29) has the same dependence on c as has that part of Eq. (5.31) for which $g = d$. So we could construct an operator

$$M = \sum_g R_g \mathbf{G}^g \cdot \mathbf{H}^g$$

whose first-order effects within $|abc\gamma\rangle$ are the same as the second-order effects of $\mathbf{D}^d \cdot \mathbf{E}^d$. In fact, to construct such an operator, we have merely to choose the reduced matrix elements of G^g_η and H^g_θ to satisfy

$$R_g\langle a \| G^g \| a\rangle\langle b \| H^g \| b\rangle = \sum_n Q_n (-1)^{a+b+a'+b'+1} \lambda(g) W\begin{pmatrix} g & b & b \\ b' & d & d \end{pmatrix} W\begin{pmatrix} g & d & d \\ a' & a & a \end{pmatrix} \tag{5.32}$$

This kind of mimicking operator is called an operator equivalent. In particular, for $g = d$ it actually mimics the first-order effects of our original operator $\mathbf{D}^d \cdot \mathbf{E}^d$ (see also Section 11.3).

CHAPTER 6

Two-Electron Formulae for the Octahedral Group

6.1 TWO-ELECTRON FUNCTIONS

We now pursue, for the octahedral group, the discussion of two-electron systems begun in Section 2.10. A one-electron operator for an n-electron system is an operator U satisfying

$$U = \sum_{i=1}^{n} u(i) \tag{6.1}$$

where $u(i)$ is a particular operator u operating on the variables for the ith electron alone. u is the same for each electron, so any permutation of the arguments i leaves U unchanged. In this chapter we shall show that, if we use a rather natural choice of basic states for a two-electron system in octahedral symmetry, we can express the matrix of a one-electron operator simply in terms of V and W coefficients (or sometimes higher ones) and the reduced matrix elements of u for the constituent one-electron orbitals.

Suppose, then, that we have a set of one-electron kets $|\frac{1}{2}m_s\rangle \mid a\alpha\rangle$ for the system, which are each products of a spin part $|\frac{1}{2}m_s\rangle$ and a spatial part $|a\alpha\rangle$. We define our two-electron states by coupling together pairs of one-electron kets, using the octahedral group coupling coefficients for the spatial part and Wigner's formula for the spin part. Then we can write the resulting two-electron states also as products of a spin part and a space part; thus

$$|a^2ShM\theta\rangle = |SM\rangle| \, a^2h\theta\rangle \tag{6.2}$$

where $|SM\rangle$ are the $\chi(SM)$ given in Eq. (2.42) and

$$\begin{aligned}|a^2h\theta\rangle &= \sum_{\alpha\beta} \langle a a \alpha\beta \mid a a h\theta\rangle| \, a\alpha(1)\rangle| \, a\beta(2)\rangle \\ &= \lambda(h)^{1/2} \sum_{\alpha\beta} V\begin{pmatrix} a & a & h \\ \alpha & \beta & \theta \end{pmatrix} | \, a\alpha(1)\rangle| \, a\beta(2)\rangle\end{aligned} \tag{6.3}$$

Only those states $|a^2ShM\theta\rangle$ for which the permutation P_{12} of the arguments of the two electrons has the eigenvalue -1 are allowed by the Pauli exclusion principle. It follows from Eq. (2.42) that

$$P_{12} \mid SM\rangle = (-1)^{S+1} \mid SM\rangle \tag{6.4}$$

and from Eq. (6.3) that

$$P_{12} \mid a^2h\theta\rangle = (-1)^h \mid a^2h\theta\rangle \tag{6.5}$$

where we have used the known behaviour of the V coefficient under interchange of its first two columns. Hence, we have shown that those states of a^2 which are allowed are those for which

$$(-1)^{h+S} = 1 \tag{6.6}$$

Equation 6.6 gives a slightly deeper significance to the fact that in two-electron configurations a^2 the triplets are always A_2 or T_1 and the singlets always A_1, E, or T_2.

Next, suppose that we have a second set $|\frac{1}{2}m_s\rangle \mid b\beta\rangle$ and that we build up the states of the configuration ab according to the same prescription. This time there are an allowed triplet and an allowed singlet for each representation contained in ab; an argument similar to that in the preceding paragraph shows that the two-electron states are

$$|abShM\theta\rangle = |SM\rangle| \, abh\theta\rangle \tag{6.7}$$

with $|SM\rangle$ as before and

$$|abh\theta\rangle = \frac{1}{\sqrt{2}}\lambda(h)^{1/2}\sum_{\alpha\beta} V\begin{pmatrix} a & b & h \\ \alpha & \beta & \theta \end{pmatrix} \{|a\alpha(1)\rangle| b\beta(2)\rangle + (-1)^S \,|\, b\beta(1)\rangle| a\alpha(2)\rangle\} \tag{6.8}$$

The two-electron states shown in Eq. (6.2) and (6.7) contain spin functions quantized according to their M_s values but not coupled in any way with the space functions. It is usual to say that they are basic states in an $S\Gamma M_s M_\Gamma$ coupling scheme. However, for two-electron states, as we remarked in Section 2.10, the set of $2S + 1$ spin functions with given S and the set of $\lambda(\Gamma)$ space functions with given Γ each separately span irreducible representations of the octahedral group O. Therefore, we may couple the spin and space functions to form bases for irreducible representations of the group O, supposing O now to operate simultaneously on the spin and spatial parts of operators and functions. To be precise, we take the real tetragonal component system $|S\sigma\rangle$ associated with the $|SM\rangle$ according to the transformation given by Eq. (2.29), and then we define our new coupled states by the equation

$$\begin{aligned} |XSht\tau\rangle &= \sum_{\theta\sigma} \langle Sh\sigma\theta \mid Sht\tau\rangle| XSh\sigma\theta\rangle \\ &= \lambda(t)^{1/2}\sum_{\sigma\theta} V\begin{pmatrix} S & h & t \\ \sigma & \theta & \tau \end{pmatrix} \mid XSh\sigma\theta\rangle \end{aligned} \tag{6.9}$$

where X stands for either a^2 or ab, and S is interpreted as 0 or 1 or as A_1 or T_1 depending on its context. The kets in Eq. (6.9) give a special case of what has previously been called an $S\Gamma\Gamma_1 M_1$ coupling scheme (24).

6.2 A PRELIMINARY LEMMA

Before we continue, it is desirable that we possess a simple proposition which simplifies the calculation of matrix elements of the quantity U of Eq. (6.1) between fully antisymmetric states, $|Z\rangle$ and $|Z'\rangle$, say, of an n-electron system. Take the transposition P_{ij}, with $i \neq j$, which interchanges the argument i with the argument j. Then P_{ij} multiplies both $|Z\rangle$ and $|Z'\rangle$ by -1. Hence,

$$\langle Z \mid u(i) \mid Z'\rangle = P_{ij}\langle Z \mid u(i) \mid Z'\rangle = \langle Z \mid u(j) \mid Z'\rangle$$

where the first equality holds because the matrix element is just a number and cannot depend on the order in which we choose to enumerate the arguments. It follows that

$$\langle Z \mid U \mid Z'\rangle = n\langle Z \mid u(i) \mid Z'\rangle \tag{6.10}$$

which is our lemma.

6.3 THE ELIMINATION OF SPIN

We shall derive general formulae for three special kinds of matrix element of a one-electron operator transforming as a component of an irreducible representation. Each of these is simply related to a matrix element for the spatial parts $|Xh\theta\rangle$ of our two-electron states, the latter being essentially the same element in each case. Accordingly, we first eliminate the spin functions in each case and then, in Section 6.4, simplify these spatial matrix elements.

The first kind of matrix element is that of a spin-independent operator U^f_ϕ in an $S\Gamma M_s M_\Gamma$ scheme. Here we have

$$\begin{aligned}\langle XShM\theta \mid U^f_\phi \mid X'S'h'M'\theta'\rangle &= 2\langle XShM\theta \mid u^f_\phi(1) \mid X'S'h'M'\theta'\rangle \\ &= 2\delta_{SS'}\delta_{MM'}\langle Xh\theta \mid u^f_\phi(1) \mid X'h'\theta'\rangle\end{aligned} \tag{6.11}$$

Such an equation may be rewritten, equivalently, in terms of reduced matrix elements for the spatial part as

$$\langle XShM \,\|\, U^f \,\|\, X'S'h'M'\rangle = 2\delta_{SS'}\delta_{MM'}\langle Xh \,\|\, u^f(1) \,\|\, X'h'\rangle \tag{6.12}$$

Next we may need a matrix element of a spin-independent operator in an $S\Gamma\Gamma_1 M_1$ scheme. This element is deduced immediately from Eq. (5.26) by putting $a = S$, $a' = S'$ and $E^e = u^f(1)$ to give us

$$\begin{aligned}\langle XSht \,\|\, U^f \,\|\, X'S'h't'\rangle = {}& 2\lambda(t)^{1/2}\lambda(t')^{1/2}(-1)^{S+h'+t+f}\delta_{SS'} \\ & \langle Xh \,\|\, u^f(1) \,\|\, X'h'\rangle W\begin{pmatrix} h' & h & f \\ t & t' & S\end{pmatrix}\end{aligned} \tag{6.13}$$

Note that we have used our lemma and also that there is no ambiguity about $(-1)^S$, although we might interpret S as either the spin or the corresponding octahedral group representation symbol.

Our third important special case is furnished by the spin-orbit coupling energy

$$\mathcal{H}_s = \sum_{i=1}^{n} \mathbf{s}(i) \cdot \mathbf{u}(i) \tag{6.14}$$

where $n = 2$ at the moment and $\mathbf{u}$ transforms as T_1. In the $S\Gamma\Gamma_1 M_1$ scheme, $\mathcal{H}_s$ is diagonal in Γ_1 and M_1, so it suffices to calculate the diagonal elements. Using Eq. (5.24) and the reduced matrix elements given by Eq. (2.45) for the spin, we find

$$\begin{aligned}&\langle Sht\tau \mid \mathcal{H}_s \mid S'h't\tau\rangle \\ &= i\sqrt{3}\,(S+S')^{1/2}(-1)^{S+S'+h+t}\langle Xh \,\|\, u(1) \,\|\, X'h'\rangle W\begin{pmatrix} h' & h & T_1 \\ S & S' & t\end{pmatrix}\end{aligned} \tag{6.15}$$

The spatial matrix element differs from the preceding ones only because here $f = T_1$, so it is less general. In each case, therefore, our calculation is complete but for the evaluation of reduced matrix elements of the type

$$\langle Xh \,\|\, u^f(1) \,\|\, X'h' \rangle$$

6.4 SIMPLIFICATION OF THE SPATIAL MATRIX ELEMENT

We must distinguish a number of cases according to the composition of the configurations from which X and X' come. First we remark that a matrix element of a one-electron operator between states of two configurations is necessarily zero unless those two configurations are either the same or differ in only one of the constituent types of one-electron function. Thus a^2 has no non-zero matrix elements of such operators with b^2 if a and b differ. This condition means that we have just four genuinely distinct cases in which a matrix element may be non-zero, namely, $(X, X') = (a^2, a^2)$, (a^2, ab), (ab, ab), and (ab, ac). Each of these must be worked out separately; we use Eq. (6.3) or Eq. (6.8) for $|Xh\theta\rangle$ and $|X'h'\theta'\rangle$.

In order to proceed, we slightly rewrite the two-electron states [Eqs. (6.3) and (6.8)] in order to be able to use Eqs. (5.25). We have

$$\begin{aligned} |a^2h\theta\rangle &= |a(1)a(2)h\theta\rangle \\ |abh\theta\rangle &= \frac{1}{\sqrt{2}}\,|\,a(1)b(2)h\theta\rangle + \frac{1}{\sqrt{2}}\,(-1)^{S+h+a+b}\,|\,b(1)a(2)h\theta\rangle \end{aligned} \tag{6.16}$$

as one readily verifies. The kets on the right-hand side are each coupled in the way that the kets are in Eq. (5.25). So we find

$$\begin{aligned} \langle a^2h \,\|\, u^f(1) \,\|\, a^2h' \rangle &= \langle a(1)a(2)h \,\|\, u^f(1) \,\|\, a(1)a(2)h' \rangle \\ &= (-1)^{h'+f}\lambda(h)^{1/2}\lambda(h')^{1/2}\langle a \,\|\, u^f \,\|\, a\rangle W\begin{pmatrix} h' & h & f \\ a & a & a \end{pmatrix} \end{aligned} \tag{6.17}$$

$$\begin{aligned} \langle a^2h \,\|\, u^f(1) \,\|\, abh' \rangle & \\ &= \frac{1}{\sqrt{2}}\,(-1)^{S'+h'+a+b}\langle a(1)a(2)h \,\|\, u^f(1) \,\|\, b(1)a(2)h' \rangle \qquad (6.18) \\ &= \frac{1}{\sqrt{2}}\,(-1)^{S'+a+b+f}\lambda(h)^{1/2}\lambda(h')^{1/2}\langle a \,\|\, u^f \,\|\, b\rangle W\begin{pmatrix} h' & h & f \\ a & b & a \end{pmatrix} \end{aligned}$$

where in deriving Eq. (6.18) one matrix element disappeared because of the delta function occurring in Eq. (5.25). The remaining two cases are quite as straightforward, and we obtain*

* The last formula given in Griffith (25), p. 461, Eq. 15, is incorrect by a factor of $(-1)^{S+S'}$.

$$\langle abh \,||\, u^f(1) \,||\, abh'\rangle = \tfrac{1}{2}(-1)^{h'+a+b+f}\lambda(h)^{1/2}\lambda(h')^{1/2}$$
$$\left[\langle a \,||\, u^f \,||\, a\rangle W\begin{pmatrix} h' & h & f \\ a & a & b\end{pmatrix} + (-1)^{S+S'+h+h'}\langle b \,||\, u^f \,||\, b\rangle W\begin{pmatrix} h' & h & f \\ b & b & a\end{pmatrix}\right] \tag{6.19}$$

$$\langle abh \,||\, u^f(1) \,||\, ach'\rangle$$
$$= \tfrac{1}{2}(-1)^{S+S'+h+a+c+f}\lambda(h)^{1/2}\lambda(h')^{1/2}\langle b \,||\, u^f \,||\, c\rangle W\begin{pmatrix} h' & h & f \\ b & c & a\end{pmatrix} \tag{6.20}$$

Sometimes we wish to know matrix elements between states for which the relative order of coupling is different from that in Eqs. (6.17)–(6.20), for example, between a^2 and ba. Such cases are dealt with by observing that

$$|baShM\theta\rangle = (-1)^{S+h+a+b} \,|\, abShM\theta\rangle \tag{6.21}$$

which follows at once from Eq. (6.8).

When we evaluate matrix elements of the spin-orbit coupling energy, we have $f = T_1$; if the one-electron functions are actually atomic orbitals, we have also $\mathbf{u} = \zeta\mathbf{l}$, where ζ is the spin-orbit coupling constant for the orbital concerned, and the reduced matrix elements of $\mathbf{l}$ for p, d, and f orbitals are given in Table 2.5. Hence, we have completely calculated the spin-orbit coupling matrix for any set of configurations based on atomic orbitals with $l \leqslant 3$. If our basic orbitals are more general molecular orbitals, the values of the reduced matrix elements of $\mathbf{u}$ will change somewhat; otherwise, everything remains the same. For spin-independent operators, our calculation is also as complete as it can be without knowing the exact nature of u^f_ϕ.

6.5 OTHER OPERATORS

More complicated operators may be treated by the same method, but then the results are usually more simply expressed in terms of X coefficients (Chapter 8). We discuss this briefly for an operator

$$O_1 = \sum_{i=1}^{2} (s(i) \times u^f(i))^{T_1}_\phi \tag{6.22}$$

which, in view of Eq. (5.15), occurs in the nuclear hyperfine interaction. Here $\mathbf{s}(i)$ is the spin vector for the ith electron. It follows from Eq. (5.23), together with the expression [Eq. (2.45)] for the reduced matrix elements of $s(1)$, that

$$\langle Sht \,||\, O_1 \,||\, S'h't'\rangle$$
$$= 3i(-1)^{S+1}(S+S')^{1/2}\lambda(t)^{1/2}\lambda(t')^{1/2}\langle h \,||\, u^f(1) \,||\, h'\rangle X\begin{bmatrix} S & h & t \\ S' & h' & t' \\ T_1 & f & T_1\end{bmatrix} \tag{6.23}$$

with the same kind of spatial matrix element $\langle Yh \,\|\, u^f(1) \,\|\, Y'h' \rangle$ that we considered previously.

The reader will now have no difficulty in dealing with any similar operator in which he may be interested, so we are ready to leave the theory of two-electron systems. First, however, we consider some examples.

6.6 g VALUES FOR THE NICKEL ION

In regular octahedral coordination, the ground term of the nickel ion is $d^8 t_2^6 e^2\ {}^3A_2$. In the strong-field coupling scheme, the first-order modification to the wave-function caused by the spin-orbit coupling arises from $t_2^5 e^3\ {}^3T_2$ and $t_2^5 e^3\ {}^1T_2$. Only the former state can influence the g values to the first order, because the matrix elements of $\mathbf{L} + 2\mathbf{S}$ between a triplet and a singlet inevitably vanish. The ${}^3T_2$ lies Δ above ${}^3A_2$, where Δ is the ligand-field splitting parameter for the d orbitals. We now calculate the g values correct to the first order.

The total ground state, spin and space, is triply degenerate and transforms as T_2; we must evaluate $\beta\mathbf{H} \cdot (\mathbf{L} + 2\mathbf{S})$ within its three components. A typical matrix element is

$$\begin{aligned} \langle T_2 i \mid \beta\mathbf{H} \cdot (\mathbf{L} + 2\mathbf{S}) \mid T_2 j \rangle \\ &= \sum_p \beta H_p \langle T_2 \,\|\, (L + 2S) \,\|\, T_2 \rangle V \begin{pmatrix} T_2 & T_2 & T_1 \\ i & j & p \end{pmatrix} \\ &= \sum_p \beta H_p \langle T_2 \,\|\, (L + 2S) \,\|\, T_2 \rangle V \begin{pmatrix} T_1 & T_1 & T_1 \\ i & j & p \end{pmatrix} \qquad (6.24) \\ &= -\frac{i}{\sqrt{6}} \langle T_2 \,\|\, (L + 2S) \,\|\, T_2 \rangle \langle 1i \mid \beta\mathbf{H} \cdot \mathbf{S} \mid 1j \rangle \end{aligned}$$

where the $|1i\rangle$ are a set of kets for $S = 1$. Consequently, the first-order interaction with an external magnetic field may be represented in a spin-Hamiltonian by $g\beta\mathbf{H} \cdot \mathbf{S}$, where

$$g = -\frac{i}{\sqrt{6}} \langle T_2 \,\|\, (L + 2S) \,\|\, T_2 \rangle \qquad (6.25)$$

We now recall [(26), Section 9.7] that matrix elements for $(t_2e)^8$ may be evaluated for the corresponding terms of $(t_2e)^2$, provided that we change the sign of such one-electron parameters as ζ. Because the only two terms which contribute to Eq. (6.25) are the triplets $e^2\ {}^3A_2$ and $t_2e\ {}^3T_2$, we deduce from the first-order perturbation expansion of $|T_2\rangle$ and the usual properties of matrix elements of $\mathbf{L}$ and $\mathbf{S}$ that

$$g = 2 - \frac{i}{\sqrt{6}} \langle T_2 \| L \| T_2 \rangle$$
$$= 2 - \frac{2i}{\Delta\sqrt{6}} \langle e^2\,{}^3A_2T_2 \| L \| et_2\,{}^3T_2T_2 \rangle \langle et_2\,{}^3T_2T_2\gamma \mid \mathcal{H}_s \mid e^2\,{}^3A_2T_2\gamma \rangle \tag{6.26}$$

where we have inverted the order of t_2 and e to facilitate application of Eq. (6.18).

The two matrix elements in Eq. (6.26) satisfy

$$\langle e^2\,{}^3A_2T_2 \| L \| et_2\,{}^3T_2T_2 \rangle = -2M$$
$$\langle et_2\,{}^3T_2T_2\gamma \| \mathcal{H}_s \| e^2\,{}^3A_2T_2\gamma \rangle = -\tfrac{1}{3}i\zeta\overline{M}\sqrt{6} \tag{6.27}$$

when we use, respectively, Eqs. (6.13) and (6.15), where

$$M = \langle e^2A_2 \| l(1) \| et_2T_2 \rangle$$

and $|M|^2 = 3$ from Eq. (6.18) and the reduced matrix element of $\mathbf{l}$ given in Table 2.5. Substituting in Eq. (6.26), we find

$$g = 2 + \frac{4\,|\,M\,|^2\,\zeta}{3\Delta} = 2 + \frac{4\zeta}{\Delta} \tag{6.28}$$

6.7 PARAMAGNETIC SUSCEPTIBILITIES

It was shown in Griffith (26), Section 10.2.1, that the matrix of spin-orbit coupling $\mathcal{H}_s$ and magnetic interaction $\beta\mathbf{H} \cdot (\mathbf{L} + 2\mathbf{S})$ within a strong-field term ${}^3T_i$ ($i = 1$ or 2) is the same as $\nu\mathbf{L} \cdot \mathbf{S}$ and $\beta\mathbf{H} \cdot (\gamma\mathbf{L} + 2\mathbf{S})$, respectively, within a ${}^3P$ atomic term. The paramagnetic susceptibility of any such ${}^3T_i$ term was then calculated in terms of the two parameters ν and γ. Hence, if we give formulae for ν and γ for our two-electron systems, we shall, in effect, have calculated the susceptibilities.

First, we calculate ν. In the ${}^3P$ term, the spin-orbit energy of ${}^3P_0$ is -2ν. The corresponding state of ${}^3T_i$ is A_i, and for it Eq. (6.15) gives

$$\langle {}^3T_iA_i \mid \mathcal{H}_s \mid {}^3T_iA_i \rangle = \tfrac{1}{3}i\zeta\sqrt{6}\,\langle XT_i \| l(1) \| XT_i \rangle$$

whence

$$\nu = -\frac{i\zeta}{\sqrt{6}} \langle XT_i \| l(1) \| XT_i \rangle \tag{6.29}$$

Similarly,

$$\gamma = \langle {}^3T_iM1 \mid L_z \mid {}^3T_iM1 \rangle$$
$$= -i\langle {}^3T_iM1 \mid L_o \mid {}^3T_iM1 \rangle$$
$$= -\frac{i}{\sqrt{6}} \langle {}^3T_iM \| L \| {}^3T_iM \rangle \tag{6.30}$$
$$= -\tfrac{1}{3}i\sqrt{6}\,\langle XT_i \| l(1) \| XT_i \rangle$$

where we used successively Eqs. (2.29), (2.34), and (6.11). Evidently, for two-electron systems in this coupling scheme we have the relation $2\nu = \gamma\zeta$. The matrix element of $l(1)$ is now evaluated in particular cases; Eqs. (6.17) and (6.19) are used.

A typical example is furnished by the $^3T_1$ term of the a_2t_2 configuration for a pair of atomic f electrons. Substitution in Eq. (6.19) yields

$$\langle XT_1 \| l(1) \| XT_1 \rangle = \tfrac{1}{4}i\sqrt{6}$$

where we also used Table 2.5. So $\gamma = \frac{1}{2}$ and $\nu = \zeta/4$. Further substitution in formulae 10.12–10.14 of Griffith (26) then gives the paramagnetic susceptibility

$$\chi = \frac{N\beta^2}{kT} \cdot \frac{3 + \frac{1}{8}(25y - 9)e^{-y} + \frac{1}{8}(125y - 15)e^{-3y}}{y(1 + 3e^{-y} + 5e^{-3y})} \tag{6.31}$$

with $y = \zeta/4kT$, in disagreement with a formula given by Griffith and Orgel.*

6.8 AN AXIAL LIGAND FIELD

As a last example, consider an axial ligand field along OZ, or, more generally, one having D_4 symmetry with the Z axis as fourfold axis and the OX, OY axes as twofold axes. Suppose that we have a two-electron system with one electron in each of a t_2 and an e orbital, and ask how the $^3T_1$ and $^3T_2$ terms of t_2e are split by the field as a function of the splitting of the e orbital, in the absence of spin-orbit coupling or any effect on the t_2 orbitals.

First, we ask, in general, how to introduce such a ligand field into our scheme of calculation. We do so by decomposing it into a sum of components of irreducible representations under O. Since it belongs to the unit representation of D_4, it can be written as a sum of a part forming a basis for A_1 of O and another part transforming as the θ component of E for O [see (26), Table A17]. The part transforming as A_1 produces a uniform shift in the position of all states of O, so, as far as the splitting is concerned, we need consider only a field V_θ^E. The only possible non-zero matrix elements of this field within the e orbitals are

$$\begin{aligned} \langle e\theta \mid V_\theta^E \mid e\theta \rangle &= -\tfrac{1}{2}\langle e \| V^E \| e \rangle \\ \langle e\epsilon \mid V_\theta^E \mid e\epsilon \rangle &= \tfrac{1}{2}\langle e \| V^E \| e \rangle \end{aligned} \tag{6.32}$$

so if we write μ for the splitting between θ and ϵ induced by V_θ^E, we have

$$\langle e \| V^E \| e \rangle = \mu \tag{6.33}$$

* *J. Chem. Phys., 26,* 988 (1957). There was a numerical error in the calculation of one of the matrix elements in that paper.

Similarly, V_θ^E splits the t_2 orbitals by δ, with t_2z lowest for δ positive, if

$$\langle t_2 \| V^E \| t_2 \rangle = \frac{2}{\sqrt{3}}\,\delta \tag{6.34}$$

We now proceed with our example and set $\delta = 0$. Then Eqs. (6.12) and (6.19) give for the t_2e configuration:

$$\begin{aligned}\langle {}^3T_iM \| \sum_{\kappa=1}^{2} V^E(\kappa) \| {}^3T_jM \rangle &= 2\langle T_i \| V^E(1) \| T_j \rangle \\ &= 3(-1)^{T_i} \langle e \| V^E \| e \rangle W \begin{pmatrix} T_j & T_i & E \\ E & E & T_2 \end{pmatrix} \\ &= \tfrac{1}{2}\mu(-1)^{T_i}\sqrt{3}\end{aligned} \tag{6.35}$$

because the W coefficient is $1/2\sqrt{3}$ for all i, j. Hence,

$$\langle {}^3T_iM\alpha \,|\, \sum_{\kappa=1}^{2} V_\theta^E(\kappa) \,|\, {}^3T_jM\beta \rangle = \tfrac{1}{2}\mu(-1)^{T_i}\sqrt{3}\; V \begin{pmatrix} T_i & T_j & E \\ \alpha & \beta & \theta \end{pmatrix} \tag{6.36}$$

where actually the matrix elements are zero unless $\alpha = \beta$.

When the field is small compared with the initial separation between ${}^3T_1$ and ${}^3T_2$, we can deduce from Eq. (6.36) that the energies of the $\alpha = \beta = x$ or y components of ${}^3T_i$ are $\frac{1}{4}(-1)^{T_i}\mu$ and of the $\alpha = \beta = z$ components is $-\frac{1}{2}(-1)^{T_i}\mu$. So the terms each split by $3\mu/4$, under this perturbation, in opposite directions.

CHAPTER 7

Fractional Parentage

7.1 COEFFICIENTS OF FRACTIONAL PARENTAGE

The remainder of the book is devoted to n-electron systems. In view of the preceding treatment of two-electron systems, we shall have especially in mind applications in which $n > 2$, although our formulae apply equally for $n = 1$ or 2. In this chapter and the following two chapters, we investigate various essential preliminary matters.

We now recall that a general coefficient of fractional parentage is defined by the equation

$$|a^nShM\theta\rangle = \sum_{S'h'S''h''} \langle a^p(S'h')a^q(S''h'')|\} a^nSh\rangle |a^p(S'h') \cdot a^q(S''h'')ShM\theta\rangle \tag{7.1}$$

where a is an irreducible representation and there is no antisymmetrization across the dot. The first p electrons are in a^p, and the last q are in a^q; i.e.,

$$|a^p(S'h') \cdot a^q(S''h'')ShM\theta\rangle = \sum_{M'M''\theta'\theta''} \langle S'S''M'M''|S'S''SM\rangle$$

$$\langle h'h''\theta'\theta'' \mid h'h''h\theta\rangle |a^pS'h'M'\theta'\rangle |a^qS''h''M''\theta''\rangle \quad (7.2)$$

with the first ket $|a^pS'h'M'\theta'\rangle$ simply the ordinary fully antisymmetric ket for the first p electrons and $|a^qS''h''M''\theta''\rangle$ similarly for electrons $p + 1$, $p + 2, \ldots, n$ in that order. The relation of the second ket to the ket $|a^qS''h''M''\theta''\rangle$ for electrons $1, 2, \ldots, q$ is simply that the argument i in the latter ket has been replaced with $p + i$ throughout.

The coefficients of fractional parentage are essential to us because Eq. (7.1) plays the same role for n-electron systems as the first of Eqs. (6.16) did for two-electron systems. In fact, the latter equation may be rewritten

$$|a^2h\theta\rangle = |a \cdot ah\theta\rangle$$

with the dot notation. We remarked in connection with Eqs. (2.42) that the spin functions used for two-electron systems satisfied

$$|SM\rangle = \sum_{m'm''} \langle \tfrac{1}{2}\tfrac{1}{2}m'm'' \mid \tfrac{1}{2}\tfrac{1}{2}SM\rangle |\tfrac{1}{2}m'\rangle |\tfrac{1}{2}m''\rangle$$

so Eq. (6.2) becomes

$$|a^2ShM\theta\rangle = |a(\tfrac{1}{2}a) \cdot a(\tfrac{1}{2}a)ShM\theta\rangle$$

in analogy with the left-hand side of Eq. (7.2). Therefore, according to the definition stated by Eq. (7.1), we have

$$\langle a, a|\}\, a^2Sh\rangle = 1 \quad (7.3)$$

whenever ${}^{2S+1}h$ is an allowed term of a^2. The $S'h'$ and $S''h''$ which strictly should occur in the coefficient have been omitted, because they are obviously redundant.

Interchanging the position of $a^p(S'h')$ and $a^q(S''h'')$ in Eq. (7.1) can only change the phase of the coefficient of fractional parentage. The phase can change, both because the order of coupling of S' to S'' and of h' to h'' is inverted and because the arguments of the electrons in a^p and a^q have to be altered. The former phase change is $(-1)^{S'+S''-S+h'+h''-h}$ (see Sections 2.5, 3.4, and 9.1). The latter phase change is achieved by the permutation $1 \longrightarrow q + 1$, $2 \longrightarrow q + 2$, etc., which is the qth power of the cycle $(1\,2\,3 \ldots n)$. This permutation multiplies $|a^nShM\theta\rangle$ by $(-1)^{(n-1)q} = (-1)^{pq}$, so we have found that

$$\langle a^q(S''h'')a^p(S'h')|\}\, a^nSh\rangle$$
$$= (-1)^{pq+S'+S''-S+h'+h''-h}\langle a^p(S'h')a^q(S''h'')|\}\, a^nSh\rangle \quad (7.4)$$

We shall use coefficients of fractional parentage only for the special cases $p = 1$ or $q = 1$. Equation (7.3) gives a general formula for them when $n = 2$. A consideration of the relationship between a p-electron system a^p and the complementary $a^{2\lambda(a)-p}$ electron system shows [(26), p. 259] that

$$\langle a^{2\lambda-1}, a|\} a^{2\lambda}\rangle = \langle a, a^{2\lambda-1}|\} a^{2\lambda}\rangle = 1 \tag{7.5}$$

There is only one term in $a^{2\lambda-1}$, namely ${}^2a$, so the sign in Eq. (7.5) is a consequence solely of the definition of the phase relation between complementary states adopted in (26), Section 9.7. We retain that phase convention in the present book.

There is a general formula for another special case, namely

$$\langle a^{2\lambda-2}(S\Gamma), a|\} a^{2\lambda-1}\rangle = \left[\frac{(2S+1)\lambda(\Gamma)}{\lambda(2\lambda-1)}\right]^{1/2} \mu(a^2 s\Gamma) \tag{7.6}$$

where $\lambda = \lambda(a)$ and $\mu = 1$ unless $\lambda = 2$. When $\lambda = 2$, then $\mu = +1$ for ${}^1A_1$ and $\mu = -1$ otherwise (here the group may be O or a dihedral group). Equation (7.6) may be proved by the method explained by Racah (15), Section 4, but we shall not do so here. Equations (7.3), (7.5), and (7.6) give all the coefficients for representations a of degree not greater than two. In particular, they deal with all the dihedral groups and leave only $n = 3, 4$ for $a = T_1$ or T_2 of the octahedral group O. The coefficients of fractional parentage for this case were first calculated by Tanabe and Sugano (20) and are given in Table 7.1.

In the rest of the chapter we discuss some examples. These are largely to illustrate the use of the coefficients and may be regarded as being special cases of the more general treatment given in Chapter 10.

7.2 MATRIX ELEMENTS OF A SPIN-INDEPENDENT OPERATOR WITHIN a^n

We now consider simplifying a matrix element like

$$E = \langle a^n ShM\theta \mid U_\phi^f \mid a^n S'h'M'\theta'\rangle \tag{7.7}$$

where U_ϕ^f is a spin-independent one-electron operator. We simplify in a number of stages, and we first remark that

$$E = \overline{E} V \begin{pmatrix} h & h' & f \\ \theta & \theta' & \phi \end{pmatrix} \delta_{SS'}\delta_{MM'} \tag{7.8}$$

where

$$\overline{E} = \langle a^n ShM \,\|\, U^f \,\|\, a^n Sh'M\rangle \tag{7.9}$$

is independent of the spin quantum number M. Hence, we shall consider only matrix elements E for which $S' = S$ and $M' = M$.

Table 7.1. Coefficients of fractional parentage for t_1^n and t_2^n configurations. Upper sign t_1, lower, t_2.

t_2^2 \ t_1^2	t_1^2 or t_2^2			
t_2^3 \ t_1^3	$^1A_1$	^1E	$^3T_1$	$^1T_2$
$^4A_2$ $^4A_1$	$\cdots$	$\cdots$	1	$\cdots$
^2E ^2E	$\cdots$	$\cdots$	$\pm\dfrac{1}{\sqrt{2}}$	$\mp\dfrac{1}{\sqrt{2}}$
$^2T_2$ $^2T_1$	$\dfrac{\sqrt{2}}{3}$	$-\tfrac{1}{3}$	$-\dfrac{1}{\sqrt{2}}$	$-\dfrac{1}{\sqrt{6}}$
$^2T_1$ $^2T_2$	$\cdots$	$\mp\dfrac{1}{\sqrt{3}}$	$\dfrac{1}{\sqrt{2}}$	$\dfrac{1}{\sqrt{6}}$

t_2^3 \ t_1^3	$^4A_2$	^2E	$^2T_2$	$^2T_1$
t_2^4 or t_1^4	$^4A_1$	^2E	$^2T_1$	$^2T_2$
$^1A_1$	$\cdots$	$\cdots$	1	$\cdots$
^1E	$\cdots$	$\cdots$	$-\tfrac{1}{2}$	$\mp\dfrac{\sqrt{3}}{2}$
$^3T_1$	$-\dfrac{1}{\sqrt{3}}$	$\dfrac{1}{\sqrt{6}}$	$-\tfrac{1}{2}$	$\tfrac{1}{2}$
$^1T_2$	$\cdots$	$-\dfrac{1}{\sqrt{2}}$	$-\tfrac{1}{2}$	$\tfrac{1}{2}$

We now expand the states of a^n by the equation

$$|a^n S h M\theta\rangle = \sum_{S_1 h_1 M_1 m} \langle a^{n-1}(S_1 h_1), a| \} a^n S h\rangle \langle S_1 \tfrac{1}{2} M_1 m \mid S_1 \tfrac{1}{2} S M\rangle$$
$$|a^{n-1}(S_1 h_1 M_1) \cdot am, h\theta\rangle \qquad (7.10)$$

In Eq. (7.10), the final ket is an irreducible product so far as space representations are concerned, but is a simple product of spin functions; i.e.,

$$|a^{n-1}(S_1 h_1 M_1) \cdot am, h\theta\rangle = \sum_{\theta_1 \alpha} \langle h_1 a\theta_1\alpha \mid h_1 a h\theta\rangle |a^{n-1} S_1 h_1 M_1 \theta_1\rangle |am\alpha\rangle$$

The somewhat unsymmetrical expansion (10) is used so that we shall be able to apply our theorems about matrix elements of irreducible products. Similarly,

$$|a^nSh'M\theta'\rangle = \sum_{S_1'h_1'M_1'm'} \langle a^{n-1}(S_1'h_1'), a|\} a^nSh'\rangle\langle S_1'\tfrac{1}{2}M_1'm' \mid S_1'\tfrac{1}{2}SM\rangle \\ |a^{n-1}(S_1'h_1'M_1') \cdot am', h'\theta'\rangle \tag{7.11}$$

On introducing the expansions given by Eqs. (7.10) and (7.11) into the matrix element E, we get a sum over $S_1h_1M_1mS_1'h_1'M_1'm'$. However, U_ϕ^f is spin-independent, and, therefore, only those terms in the sum remain for which $S_1' = S_1$, $M_1' = M_1$, $m' = m$. To make the derivation more succinct, we put

$$p = \langle a^{n-1}(S_1h_1), a|\} a^nSh\rangle\langle a^{n-1}(S_1h_1'), a|\} a^nSh'\rangle$$

and then

$$\begin{aligned}\bar{E} &= \langle a^nShM \,||\, U^f \,||\, a^nSh'm\rangle \\ &= n\langle a^nShM \,||\, u^f(n) \,||\, a^nSh'M\rangle\end{aligned}$$

by Eq. (6.10), after which

$$\bar{E} = n \sum_{S_1h_1M_1mh_1'} p|\langle S_1\tfrac{1}{2}SM \mid S_1\tfrac{1}{2}M_1m\rangle|^2 \\ \langle a^{n-1}(S_1h_1M_1) \cdot am, h \,||\, u^f(n) \,||\, a^{n-1}(S_1h_1'M_1) \cdot am, h'\rangle$$

The Wigner coefficients are the coefficients in an orthonormal transformation, and the reduced matrix elements are independent of M_1 and m, so summation over M_1 and m gives

$$\bar{E} = n \sum_{S_1h_1h_1'} p\langle a^{n-1}(S_1h_1M_1) \cdot am, h \,||\, u^f(n) \,||\, a^{n-1}(S_1h_1'M_1) \cdot am, h'\rangle$$

Here we have a weighted sum of reduced matrix elements of a very simple irreducible product; hence, using Eq. (5.26), we obtain

$$\bar{E} = \langle a \,||\, u^f \,||\, a\rangle \sum_{S_1h_1} np(-1)^{h_1+a+f+h}\lambda(h)^{1/2}\lambda(h')^{1/2}W\begin{pmatrix} a & a & f \\ h & h' & h_1\end{pmatrix} \tag{7.12}$$

where now

$$p = \langle a^{n-1}(S_1h_1), a|\} a^nSh\rangle\langle a^{n-1}(S_1h_1), a|\} a^nSh'\rangle \tag{7.13}$$

If we write $g^n_{hh'}(a, f)$ for the sum in Eq. (7.12) we have the succinct formula

$$\langle a^nShM \,||\, U^f \,||\, a^nSh'M\rangle = \langle a \,||\, u^f \,||\, a\rangle g^n_{hh'}(a, f) \tag{7.14}$$

illustrating that we have expressed the matrix of U^f for the n-electron system as a multiple of the reduced matrix elements of u^f for the corresponding one-electron system. As the notation indicates, the quantities $g^n_{hh'}(a, f)$ depend on the five parameters n, h, h', a, and f (the sixth, S, is determined implicitly by the other five). However, they are pure numbers which are entirely independent of any information about the operator U

except that it is spin-independent, is a one-electron operator, and transforms according to the irreducible representation f. For a given group, they can be tabulated once and for all. They play a central part in Tanabe and Kamimura's treatment (21) of matrix elements of operators like U^f for systems $a^m b^n$, and we shall meet them again in Chapter 10. We now derive their values.

For $n = 1$, the g^n are not really defined, but we could put $g^1 = 1$ without inconsistency if we wished. When $f = A_1$, we use Eq. (4.2), and we can then sum over $S_1 h_1$ to give

$$g^n_{hh'}(a, A_1) = n\delta_{hh'}\lambda(a)^{-1/2}\lambda(h')^{1/2}$$

whence

$$\begin{aligned}\langle a^n ShM\theta \mid U^{A_1} \mid a^n ShM\theta\rangle &= n\lambda(a)^{-1/2}\langle a \| u^{A_1} \| a\rangle \\ &= n\langle a\alpha \mid u^{A_1} \mid a\alpha\rangle\end{aligned} \tag{7.15}$$

which states simply that if a has orbital energy δ then any state of a^n has diagonal energy $n\delta$. Next, if $n = 2$, each coefficient of fractional parentage is unity for terms allowed in a^2, and there is only one term ${}^{2S_1+1}h_1$, namely ${}^2a$. Hence

$$g^2_{hh'}(a, f) = 2(-1)^{h+f}\lambda(h)^{1/2}\lambda(h')^{1/2} W\begin{pmatrix} a & a & f \\ h & h' & a\end{pmatrix} \tag{7.16}$$

If $n = 2\lambda(a)$, we have a filled shell, so $h = h_1' = A_1$, and $g = 0$ unless $f = A_1$. If $n = 2\lambda(a) - 1$, then we use Eq. (7.6) to give

$$g^{2\lambda(a)-1}_{aa}(a, f) = \sum_{S_1 h_1} (2S_1 + 1)\lambda(h_1)(-1)^{h+f} W\begin{pmatrix} a & a & f \\ a & a & h_1\end{pmatrix}$$

Surprisingly enough, the sum on the right-hand side can be simplified. We utilize the fact that $(-1)^{S_1+h_1} = 1$ always, which means that $2S_1 + 1$ can be replaced by $2 + (-1)^{h_1+1}$. Then we use Eq. (4.13), with $f = f$, $g = A_1$, and Eq. (4.15) to obtain

$$g^{2\lambda(a)-1}_{aa}(a, f) = (-1)^{f+1}\delta(a, a, f) \tag{7.17}$$

unless $f = A_1$. We have now derived general formulae for all n, h, h', a, f when $\lambda(a) = 1$ or 2.

The only values of g which remain now are when $n = 3$ or 4 for representations of degree 3. We now recall [(26), Eq. (9.38b)] that the matrix of a one-electron operator for $a^{2\lambda(a)-n}$ is $-\eta$ times that for a^n when $f \neq A_1$, $n \neq \lambda(a)$, and η is as it was defined in Section 2.6. Hence, for f contained in both the direct products hh' and a^2,

$$\langle a \| u^f \| a\rangle g^{2\lambda-n}_{hh'}(a, f) V\begin{pmatrix} h & h' & f \\ \theta & \theta' & \phi\end{pmatrix} = -\eta\langle a \| u^f \| a\rangle g^n_{hh'}(a, f) V\begin{pmatrix} h & h' & f \\ \theta & \theta' & \phi\end{pmatrix}$$

or

$$\langle a \| u^f \| a\rangle (g^{2\lambda-n}_{hh'}(a, f) + \eta g^n_{hh'}(a, f)) = 0$$

Now $\langle a \| u^f \| a\rangle = 0$ unless $\eta = (-1)^f$, when it is not necessarily zero (Section 2.6), so

$$g_{hh'}^{2\lambda-n}(a,f) = (-1)^{f+1}g_{hh'}^{n}(a,f) \tag{7.18}$$

consistently with Eq. (7.17). Consequently, for representations of degree 3, we deduce from Eq. (7.16) that

$$g_{hh'}^{4}(a,f) = 2(-1)^{h+1}\lambda(h)^{1/2}\lambda(h')^{1/2}W\begin{pmatrix} a & a & f \\ h & h' & a\end{pmatrix} \tag{7.19}$$

Finally, for $n = 3$, $a = t_1$ or t_2, the g may be calculated [Tanabe and Kamimura (21)]; they are given in Table 7.2.

Table 7.2. Values of $g_{hh'}(a,f)$ for the t_1^3 and t_2^3 configurations of the octahedral group. Upper sign for t_1, lower for t_2.

$f = A_1$	t_2^3	^2E	$^2T_2$	$^2T_1$
t_2^3	t_1^3	^2E	$^2T_1$	$^2T_2$
^2E	^2E	$\sqrt{6}$	...	...
$^2T_2$	$^2T_1$	...	3	...
$^2T_1$	$^2T_2$	...	...	3

$f = E$	t_2^3	^2E	$^2T_2$	$^2T_1$
t_2^3	t_1^3	^2E	$^2T_1$	$^2T_2$
^2E	^2E	...	...	...
$^2T_2$	$^2T_1$	...	...	$\mp\sqrt{3}$
$^2T_1$	$^2T_2$	...	$\pm\sqrt{3}$	...

$f = T_1$	t_2^3	^2E	$^2T_2$	$^2T_1$
t_2^3	t_1^3	^2E	$^2T_1$	$^2T_2$
^2E	^2E	...	...	$\sqrt{2}$
$^2T_2$	$^2T_1$	...	1	...
$^2T_1$	$^2T_2$	$\pm\sqrt{2}$	...	-1

$f = T_2$	t_2^3	^2E	$^2T_2$	$^2T_1$
t_2^3	t_1^3	^2E	$^2T_1$	$^2T_2$
^2E	^2E	...	$\pm\sqrt{2}$	...
$^2T_2$	$^2T_1$	$-\sqrt{2}$	...	1
$^2T_1$	$^2T_2$	...	-1	...

7.3 MATRIX ELEMENTS OF U_ϕ^f BETWEEN a^n AND $a^{n-1}b$

A problem very similar to the preceding one is the simplification of

$$F = \langle a^n ShM\theta \mid U_\phi^f \mid a^{n-1}(S_1'h_1')bS'h'M'\theta'\rangle \tag{7.20}$$

where U_ϕ^f is a spin-independent one-electron operator as before, and the functions b span the irreducible representation b and are orthogonal to the set of functions a. Clearly, $F = 0$ unless $S' = S$, $M' = M$, so we suppose these equalities to hold. We replace U_ϕ^f by $nu_\phi^f(n)$, according to the lemma of Section 6.2.

The expansion of $|a^n ShM\theta\rangle$ by Eq. (7.10) is used again, but in place of Eq. (7.11) we now have

$$|a^{n-1}(S_1'h_1')bSh'M\theta'\rangle = \sum_{M_1'm'} \langle S_1'\tfrac{1}{2}M_1'm' \mid S_1'\tfrac{1}{2}SM\rangle |a^{n-1}(S_1'h_1'M_1')bm', h'\theta'\rangle \tag{7.21}$$

This time, all the kets are fully antisymmetrized. However, since every ket $|am\alpha\rangle$ is orthogonal to every ket $|bm'\beta\rangle$, it follows that the only non-zero contribution to F must come from those pieces of Eq. (7.20) in which $u_\phi^f(n)$ connects one-electron kets $|am\alpha\rangle$ and $|bm'\beta\rangle$ for the nth electron. Hence, the only part of Eq. (7.21) which contributes is that in which the nth electron is in a one-electron ket $|bm'\beta\rangle$. This part is

$$|a^{n-1}(S_1'h_1')bSh'M\theta'\rangle_n = \frac{1}{\sqrt{n}} \sum_{M_1'm'} \langle S_1'\tfrac{1}{2}M_1'm' \mid S_1'\tfrac{1}{2}SM\rangle |a^{n-1}(S_1'h_1'M_1') \cdot bm', h'\theta'\rangle \tag{7.22}$$

The normalizing factor $1/\sqrt{n}$ occurs because the ket on the left-hand side of Eq. (7.21) has been broken up into n orthogonal parts, the nth of which is shown in Eq. (7.22). The general problem of determining normalizing factors for composite states of this type is discussed in Sections 10.3 and 10.4

Equation (7.22) exactly parallels Eq. (7.11) with the sole exception that the nth electron is now b, not a. Therefore, the remainder of our present calculation is practically the same as our previous treatment of the matrix element of E; we obtain for the reduced matrix element of F:

$$\begin{aligned}\bar{F} &= \sqrt{n}\,\langle a^{n-1}(S_1'h_1'), a|\}a^n Sh\rangle \\ &\qquad \langle a^{n-1}(S_1'h_1'M_1) \cdot am, h \,\|\, u^f(n) \,\|\, a^{n-1}(S_1'h_1'M_1) \cdot bm, h'\rangle \\ &= \langle a \,\|\, u^f \,\|\, b\rangle\langle a^{n-1}(S_1'h_1'), a|\}a^n Sh\rangle \\ &\qquad \sqrt{n}\,(-1)^{h_1'+b+h+f}\lambda(h)^{1/2}\lambda(h')^{1/2} W\begin{pmatrix} b & a & f \\ h & h' & h_1' \end{pmatrix}\end{aligned} \tag{7.23}$$

where we used Eq. (5.26).

We are often interested in the case in which a^n is a filled shell of a electrons; i.e., $n = 2\lambda(a)$. In that case, $h = A_1, f = h'$, and $h_1' = a$, so Eq. (4.2) gives a general formula for W and Eq. (7.5) gives the coefficient of fractional parentage, whence

$$\bar{F} = \sqrt{2}\,\langle a \,\|\, u^f \,\|\, b\rangle \tag{7.24}$$

A simple application of this result is given in the next section.

7.4 NUCLEAR MAGNETIC RESONANCE FOR THE COBALTIC ION

The interpretation of nuclear magnetic resonance measurements for low-spin cobaltic compounds was discussed previously;* it was shown that if the states of the cobaltic ion can be correctly represented by the strong-field coupling scheme, we should have

$$\sigma = -\frac{4\beta^2}{E}\,\overline{r^{-3}}\,|\langle t_2^6\,{}^1A_1 \mid L_z \mid t_2^5e\,{}^1T_{1z}\rangle|^2 \tag{7.25}$$

for the chemical screening constant, where E is the energy separation between the ${}^1A_1$ and ${}^1T_1$ terms. The square modulus of the matrix element in Eq. (7.25) was calculated directly, but we now see that it follows immediately from Eq. (7.24). We have

$$\begin{aligned} |\langle t_2^6\,{}^1A_1 \mid L_z \mid t_2^5e\,{}^1T_{1z}\rangle|^2 &= |\langle t_2^6\,{}^1A_1 \| L \| t_2^5e\,{}^1T_1\rangle|^2 V\begin{pmatrix} A_1 & T_1 & T_1 \\ \iota & z & z \end{pmatrix}^2 \\ &= \tfrac{2}{3}|\langle t_2 \| l \| e\rangle|^2 = 8 \end{aligned}$$

using Table 2.5 and assuming that our orbitals are d orbitals. Therefore,

$$\sigma = -\frac{32\beta^2}{E}\,\overline{r^{-3}}$$

* J. S. Griffith and L. E. Orgel, *Trans. Farad. Soc.*, **53**, 601 (1957).

CHAPTER 8

X Coefficients

8.1 DEFINITION AND SOME GENERAL PROPERTIES

The next independent invariant after W, in order of increasing complexity, is

$$X\begin{bmatrix} a & b & c \\ d & e & f \\ g & h & k \end{bmatrix} = \sum_{\alpha\beta\gamma\delta\epsilon\phi\eta\theta\kappa} V\begin{pmatrix} a & b & c \\ \alpha & \beta & \gamma \end{pmatrix} V\begin{pmatrix} d & e & f \\ \delta & \epsilon & \phi \end{pmatrix} V\begin{pmatrix} g & h & k \\ \eta & \theta & \kappa \end{pmatrix}$$
$$V\begin{pmatrix} a & d & g \\ \alpha & \delta & \eta \end{pmatrix} V\begin{pmatrix} b & e & h \\ \beta & \epsilon & \theta \end{pmatrix} V\begin{pmatrix} c & f & k \\ \gamma & \phi & \kappa \end{pmatrix} \quad (8.1)$$

We have already met it in our discussions of matrix elements of general irreducible products [Eqs. (5.23) and (6.23)], and it will appear again in our formulae for n-electron systems in Chapter 10. Evidently, X is zero unless

$$\delta(a, b, c)\delta(d, e, f)\delta(g, h, k)\delta(a, d, g)\delta(b, e, h)\delta(c, f, k) = 1 \tag{8.2}$$

It is apparent from its definition that X is invariant to any even permutation of its rows, its columns, or both. It is also invariant to transposition; i.e.,

$$X\begin{bmatrix} a & d & g \\ b & e & h \\ c & f & k \end{bmatrix} = X\begin{bmatrix} a & b & c \\ d & e & f \\ g & h & k \end{bmatrix} \tag{8.3}$$

In these respects, it is behaving like a 3×3 determinant. Under odd permutation of either its rows or its columns, it becomes multiplied by

$$\rho = (-1)^{a+b+c+d+e+f+g+h+k}$$

which may be $+1$ or -1, depending upon the nature of the constituent representations. If $\rho = -1$ and a pair of rows (or columns) of X are identical, we can deduce that $X = 0$ just as in the elementary theory of determinants. For example,

$$X\begin{bmatrix} A_2 & E & E \\ E & E & E \\ E & E & E \end{bmatrix} = 0$$

for a dihedral group, or the octahedral group, when E is any irreducible representation of degree 2 of the group concerned. As well as these, X possesses another and somewhat less straightforward symmetry, which is described in Appendix B.

A general X is expressible as a sum of products of W coefficients, and we derive that formula now. In order to be able to deduce, in Section 8.4, various other equations satisfied by X, we actually make our derivation for a more general quantity, which we will write $X(f, f')$. This quantity is defined by the right-hand side of Eq. (8.1), except that the second f and ϕ therein have been changed to f' and ϕ' respectively, there is no summation over ϕ or ϕ', and a factor $\lambda(f)$ has been introduced. Evidently, when $f' = f$ and $\phi' = \phi$, we have

$$X = \lambda(f)^{-1} \sum_{\phi} X(f, f) \tag{8.4}$$

We now start with the (2, 2) equation [Eq. (4.11)] for the W coefficients, with a change of notation for the representation symbols:

$$\sum_{\beta} V\begin{pmatrix} a & b & c \\ \alpha & \beta & \gamma \end{pmatrix} V\begin{pmatrix} b & e & h \\ \beta & \epsilon & \theta \end{pmatrix} = \sum_{\beta} V\begin{pmatrix} e & h & b \\ \epsilon & \theta & \beta \end{pmatrix} V\begin{pmatrix} a & b & c \\ \alpha & \beta & \gamma \end{pmatrix}$$

$$= \sum_{p\pi} \lambda(p) W\begin{pmatrix} e & a & p \\ c & h & b \end{pmatrix} V\begin{pmatrix} e & a & p \\ \epsilon & \alpha & \pi \end{pmatrix} V\begin{pmatrix} p & c & h \\ \pi & \gamma & \theta \end{pmatrix}$$

We introduce the foregoing equation into the definition of $X(f, f')$, and then use Eq. (4.10) twice to give

$$X(f,f') = \lambda(f) \sum_{p\pi\alpha\gamma\delta\epsilon\eta\theta\kappa} \lambda(p) W \begin{pmatrix} p & e & a \\ b & c & h \end{pmatrix} V \begin{pmatrix} p & c & h \\ \pi & \gamma & \theta \end{pmatrix} V \begin{pmatrix} c & f' & k \\ \gamma & \phi' & \kappa \end{pmatrix}$$
$$V \begin{pmatrix} g & h & k \\ \eta & \theta & \kappa \end{pmatrix} V \begin{pmatrix} p & e & a \\ \pi & \epsilon & \alpha \end{pmatrix} V \begin{pmatrix} a & d & g \\ \alpha & \delta & \eta \end{pmatrix} V \begin{pmatrix} d & e & f \\ \delta & \epsilon & \phi \end{pmatrix}$$
$$= \lambda(f) \sum_{p\pi\eta} \lambda(p) W \begin{pmatrix} p & e & a \\ b & c & h \end{pmatrix} W \begin{pmatrix} p & g & f' \\ k & c & h \end{pmatrix} V \begin{pmatrix} p & g & f' \\ \pi & \eta & \phi' \end{pmatrix} \tag{8.5}$$
$$W \begin{pmatrix} p & g & f \\ d & e & a \end{pmatrix} V \begin{pmatrix} p & g & f \\ \pi & \eta & \phi \end{pmatrix}$$
$$= \delta_{ff'}\delta_{\phi\phi'} \sum_{p} \lambda(p) W \begin{pmatrix} p & e & a \\ b & c & h \end{pmatrix} W \begin{pmatrix} p & c & h \\ k & g & f \end{pmatrix} W \begin{pmatrix} p & g & f \\ d & e & a \end{pmatrix}$$

where the last step used Eq. (2.18). We put $f = f'$ and $\phi = \phi'$ and substitute this result in Eq. (8.4). We obtain

$$X \begin{bmatrix} a & b & c \\ d & e & f \\ g & h & k \end{bmatrix} = \sum_{p} \lambda(p) W \begin{pmatrix} p & e & a \\ b & c & h \end{pmatrix} W \begin{pmatrix} p & c & h \\ k & g & f \end{pmatrix} W \begin{pmatrix} p & g & f \\ d & e & a \end{pmatrix} \tag{8.6}$$

and, on resubstituting from Eq. (8.6) into Eq. (8.5),

$$X(f,f') = X\delta_{ff'}\delta_{\phi\phi'} \tag{8.7}$$

Of course, there is nothing special about the representation f; we could have obtained the same result for any of the nine constituent representations of X.

8.2 SPECIAL FORMULAE FOR X

The first special formula occurs when one of the representation symbols in X is A_1. In that case, two of the V coefficients in the sum in [Eq. (8.1)] reduce according to Eq. (2.20) to multiples of products of δ functions and X reduces to a multiple of a sum of products of four V coefficients which is easily recognized as a W coefficient. For example, when $k = A_1$, we find

$$X \begin{bmatrix} a & b & c \\ d & e & f \\ g & h & A_1 \end{bmatrix} = (-1)^{b+d+f+h} \lambda(c)^{-1/2} \lambda(g)^{-1/2} \delta_{cf} \delta_{gh} W \begin{pmatrix} a & b & c \\ e & d & g \end{pmatrix} \tag{8.8}$$

All other cases can be subsumed under this one by rearranging X so that the A_1 representation is found in the bottom right-hand corner.

A second special formula follows from the Biedenharn-Elliott identity [Eq. (4.16)]. We write

$$s = a + b + c + d + e + f + g + h + k$$

and then we rewrite Eq. (4.16) in a changed notation as

$$W\begin{pmatrix} a & e & k \\ f & g & d \end{pmatrix} W\begin{pmatrix} a & e & k \\ h & c & b \end{pmatrix}$$
$$= (-1)^s \sum_p (-1)^p \lambda(p) W\begin{pmatrix} p & b & d \\ e & f & h \end{pmatrix} W\begin{pmatrix} p & h & f \\ k & g & c \end{pmatrix} W\begin{pmatrix} p & g & c \\ a & b & d \end{pmatrix} \quad (8.9)$$

where it looks rather like formula (8.6). In fact, if we first rearrange X and then apply Eq. (8.6), we get

$$X\begin{bmatrix} a & b & c \\ d & e & f \\ g & h & k \end{bmatrix} = (-1)^s X\begin{bmatrix} d & e & f \\ a & b & c \\ g & h & k \end{bmatrix}$$
$$= (-1)^s \sum_p \lambda(p) W\begin{pmatrix} p & b & d \\ e & f & h \end{pmatrix} W\begin{pmatrix} p & f & h \\ k & g & c \end{pmatrix} W\begin{pmatrix} p & g & c \\ a & b & d \end{pmatrix}$$

so if $f = h$ (or $c = g$) and the sum over p happens not to include A_2 or T_1, we can deduce that

$$X\begin{bmatrix} a & b & c \\ d & e & f \\ g & h & k \end{bmatrix} = W\begin{pmatrix} a & e & k \\ f & g & d \end{pmatrix} W\begin{pmatrix} a & e & k \\ h & c & b \end{pmatrix} \quad (8.10)$$

As an example, we have, for the octahedral group,

$$X\begin{bmatrix} E & E & E \\ E & T_1 & T_2 \\ E & T_2 & T_1 \end{bmatrix} = W\begin{pmatrix} E & T_1 & T_1 \\ T_2 & E & E \end{pmatrix}^2 = \frac{1}{12}$$

Another relation follows from

$$\sum_{c\gamma} \lambda(c) V\begin{pmatrix} a & b & c \\ \alpha & \beta & \gamma \end{pmatrix} V\begin{pmatrix} a & b & c \\ \phi & \kappa & \gamma \end{pmatrix} = \delta_{\alpha\phi}\delta_{\beta\kappa} \quad (8.11)$$

which is equivalent to one of Eqs. (2.18). We put $a = f$ and $b = k$ in Eq. (8.1) and use Eq. (8.11), whence*

$$\sum_c \lambda(c) X\begin{bmatrix} a & b & c \\ d & e & a \\ g & h & b \end{bmatrix} = \sum_{\alpha\beta\delta\epsilon\eta\theta} V\begin{pmatrix} d & e & a \\ \delta & \epsilon & \alpha \end{pmatrix} V\begin{pmatrix} g & h & b \\ \eta & \theta & \beta \end{pmatrix} V\begin{pmatrix} a & d & g \\ \alpha & \delta & \eta \end{pmatrix} V\begin{pmatrix} b & e & h \\ \beta & \epsilon & \theta \end{pmatrix}$$
$$= \sum_{\epsilon\eta} \lambda(e)^{-1}\delta_{eg}\delta_{\epsilon\eta}\delta(a, d, e)\lambda(e)^{-1}\delta_{eg}\delta_{\epsilon\eta}\delta(h, b, e) \quad (8.12)$$
$$= \lambda(e)^{-1}\delta_{eg}\delta(a, d, e)\delta(b, e, h)$$

* In Griffith (25), p. 294, Eq. (8.12) is incorrectly given in that the factor $\delta\,(a, d, e)$ is omitted.

For the octahedral group there is another special case analogous to Eq. (4.3). It is

$$X\begin{bmatrix} a & b & T_1 \\ d & e & T_1 \\ T_1 & T_1 & T_1 \end{bmatrix} = \tfrac{1}{6}\lambda(b)^{-1}\delta_{bd} - \tfrac{1}{6}\lambda(a)^{-1}\delta_{ae}(-1)^{b+d} \tag{8.13}$$

Its proof is left as an exercise for the reader; Eqs. (2.22) and (2.23) should be used.

The evaluation of the X is fairly straightforward. As well as the formulae given in this section, it is useful to employ also the extra symmetry relations derived in Appendix B. An effectively complete tabulation of X for the groups O, D_3 (and implicitly for D_6), and D_5 appears in Appendices C and D. The group D_4 is treated in the next section.

8.3 *X* COEFFICIENTS FOR THE GROUP D_4

It is interesting that essentially all the X for D_4 can be given as a few simple formulae. The value of X is zero unless it satisfies Eq. (8.2); if it contains an A_1 representation, it is easily deduced from Eq. (8.8). Among the remaining X, those which do not contain the representation E are rearrangements of

$$X\begin{bmatrix} A_2 & B_1 & B_2 \\ B_2 & A_2 & B_1 \\ B_1 & B_2 & A_2 \end{bmatrix} = -1$$

Bearing in mind that the direct product $E^2 = A_1 + A_2 + B_1 + B_2$, we may rearrange all other X coefficients into one of the two forms

$$X\begin{bmatrix} E & E & \alpha \\ E & E & \beta \\ \alpha & \beta & \gamma \end{bmatrix} \quad \text{and} \quad X\begin{bmatrix} \alpha & E & E \\ E & \beta & E \\ E & E & \gamma \end{bmatrix}$$

with each of α, β, γ of degree 1. To give general formulae for these, we note that the W [Appendix D and Eq. (4.2)] satisfy

$$W\begin{pmatrix} E & E & \alpha \\ E & E & \beta \end{pmatrix} = \tfrac{1}{2}(-1)^{\alpha\beta}, \qquad W\begin{pmatrix} \alpha & \beta & \gamma \\ E & E & E \end{pmatrix}^2 = \tfrac{1}{2}\delta(\alpha, \beta, \gamma)$$

even in the case that one or more of α, β, γ are A_1.

The first form of X may be derived from Eq. (8.10). This gives

$$X\begin{bmatrix} E & E & \alpha \\ E & E & \beta \\ \alpha & \beta & \gamma \end{bmatrix} = W\begin{pmatrix} E & E & \gamma \\ \beta & \alpha & E \end{pmatrix}^2 = W\begin{pmatrix} \alpha & \beta & \gamma \\ E & E & E \end{pmatrix}^2 = \tfrac{1}{2}\delta(\alpha, \beta, \gamma)$$

The second, from Eq. (8.6), is

$$X\begin{bmatrix}\alpha & E & E\\ E & \beta & E\\ E & E & \gamma\end{bmatrix} = \sum_p \lambda(p) W\begin{pmatrix}p & \beta & \alpha\\ E & E & E\end{pmatrix} W\begin{pmatrix}p & E & E\\ \gamma & E & E\end{pmatrix} W\begin{pmatrix}p & E & E\\ E & \beta & \alpha\end{pmatrix} = \tfrac{1}{4}(-1)^{\alpha\beta\gamma}$$

because only the term in the sum with $p = \alpha\beta$ can be non-zero.

The set of formulae we have derived for the values of X for the group D_4, is complete in the sense that any X, apart from those which are trivially zero because they do not satisfy Eq. (8.2), can be rearranged into one of them.

8.4 GENERAL EQUATIONS SATISFIED BY X COEFFICIENTS

We multiply Eq. (8.7) through by

$$\lambda(f')V\begin{pmatrix}c & f' & k\\ \gamma' & \phi' & \kappa'\end{pmatrix}$$

and sum over f' and ϕ'. Then the right-hand side becomes simply

$$\lambda(f)XV\begin{pmatrix}c & f & k\\ \gamma' & \phi & \kappa'\end{pmatrix}$$

The left-hand side simplifies somewhat because of

$$\sum_{f'\phi'} \lambda(f')V\begin{pmatrix}c & f' & k\\ \gamma' & \phi' & \kappa'\end{pmatrix} V\begin{pmatrix}c & f' & k\\ \gamma & \phi' & \kappa\end{pmatrix} = \delta_{\gamma\gamma'}\delta_{\kappa\kappa'}$$

from Eq. (2.18). Let us now drop the primes from γ' and κ' and write out what we have obtained:

$$X\begin{bmatrix}a & b & c\\ d & e & f\\ g & h & k\end{bmatrix} V\begin{pmatrix}c & f & k\\ \gamma & \phi & \kappa\end{pmatrix}$$
$$= \sum_{\alpha\beta\delta\epsilon\eta\theta} V\begin{pmatrix}a & b & c\\ \alpha & \beta & \gamma\end{pmatrix} V\begin{pmatrix}d & e & f\\ \delta & \epsilon & \phi\end{pmatrix} V\begin{pmatrix}g & h & k\\ \eta & \theta & \kappa\end{pmatrix} V\begin{pmatrix}a & d & g\\ \alpha & \delta & \eta\end{pmatrix} V\begin{pmatrix}b & e & h\\ \beta & \epsilon & \theta\end{pmatrix} \quad (8.14)$$

Equation (8.14) is very similar to Eq. (4.10) for the W. Just as for W, so for X there is a whole series of equations which can now be derived by using Eqs. (2.18) and (8.14). Since their proofs are all entirely straightforward, the more important ones are simply quoted, leaving the details for the reader.

$$\sum_{\alpha\beta\delta\epsilon} V\begin{pmatrix}a & b & c\\ \alpha & \beta & \gamma\end{pmatrix} V\begin{pmatrix}d & e & f\\ \delta & \epsilon & \phi\end{pmatrix} V\begin{pmatrix}a & d & g\\ \alpha & \delta & \eta\end{pmatrix} V\begin{pmatrix}b & e & h\\ \beta & \epsilon & \theta\end{pmatrix}$$
$$= \sum_{k\kappa} \lambda(k) X\begin{bmatrix}a & b & c\\ d & e & f\\ g & h & k\end{bmatrix} V\begin{pmatrix}c & f & k\\ \gamma & \phi & \kappa\end{pmatrix} V\begin{pmatrix}g & h & k\\ \eta & \theta & \kappa\end{pmatrix} \quad (8.15)$$

$$\sum_{\alpha\delta} V\begin{pmatrix} a & b & c \\ \alpha & \beta & \gamma \end{pmatrix} V\begin{pmatrix} d & e & f \\ \delta & \epsilon & \phi \end{pmatrix} V\begin{pmatrix} a & d & g \\ \alpha & \delta & \eta \end{pmatrix}$$

$$= \sum_{hk\theta\kappa} \lambda(h)\lambda(k) X\begin{bmatrix} a & b & c \\ d & e & f \\ g & h & k \end{bmatrix} V\begin{pmatrix} c & f & k \\ \gamma & \phi & \kappa \end{pmatrix} V\begin{pmatrix} g & h & k \\ \eta & \theta & \kappa \end{pmatrix} V\begin{pmatrix} b & e & h \\ \beta & \epsilon & \theta \end{pmatrix} \quad (8.16)$$

$$\sum_{gh} \lambda(g)\lambda(h) X\begin{bmatrix} a & b & c \\ d & e & f \\ g & h & k \end{bmatrix} X\begin{bmatrix} a & b & c' \\ d & e & f' \\ g & h & k \end{bmatrix} \quad (8.17)$$

$$= \lambda(c)^{-1}\lambda(f)^{-1}\delta_{cc'}\delta_{ff'}\delta(a, b, c)\delta(d, e, f)\delta(c, f, k)$$

$$\sum_{gh} (-1)^h \lambda(g)\lambda(h) X\begin{bmatrix} a & d & g \\ b & e & h \\ c & f & k \end{bmatrix} X\begin{bmatrix} a & d & g \\ e & b & h \\ l & m & k \end{bmatrix} = (-1)^{f+m} X\begin{bmatrix} a & b & c \\ e & d & f \\ l & m & k \end{bmatrix} \quad (8.18)$$

CHAPTER 9

Spin

9.1 REDUCED MATRIX ELEMENTS AND $\overline{V}$ COEFFICIENTS

We take our basic states for an n-electron system not merely as bases for irreducible representations of the appropriate finite symmetry group but also as eigenfunctions of the total spin, S, and the z component of spin, S_z.* Hence, we expect to need a mathematical apparatus for dealing with spin analogous to that already described for spatial functions. Such an apparatus has been given by Racah and Wigner (3, 4); in the first three sections of this chapter we recall briefly the features of it which are necessary for us.† Historically, Racah's

* Throughout, spin and orbital angular momentum vectors are in units of $\hbar$.

† For introduction to the basic quantum mechanics of spin angular momenta, see ref. (1), ref. (26), or P. A. M. Dirac, *The Principles of Quantum Mechanics*, 4th ed. Oxford, at the Clarendon Press. For more detail about $\overline{V}$, $\overline{W}$, and $\overline{X}$, see ref. (2)–(5).

and Wigner's work preceded the discussion of finite symmetry groups, and the theory for the latter was inspired by it.

Suppose now that we have eigenstates $|SM\rangle$ and $|S'M'\rangle$ of $\mathbf{S}^2$ and S_z and also a set of operators f_η^g which are operator eigenstates of $\mathbf{S}^2$ and S_z with respective eigenvalues $g(g+1)$ and η and correctly connected in phase. The property of being operator eigenstates may be expressed analytically by the requirement that the commutators of the components of $\mathbf{S}$ with the f_η^g satisfy the usual equations expected of eigenstates of $\mathbf{S}^2$, S_z. Thus

$$\begin{aligned}[S_z, f_\eta^g] &= S_z f_\eta^g - f_\eta^g S_z = \eta f_\eta^g \\ [S^\pm, f_\eta^g] &= \{(g \mp \eta)(g \pm \eta + 1)\}^{1/2} f_{\eta+1}^g\end{aligned} \tag{9.1}$$

where $S^\pm = S_x \pm iS_y$ are the shift operators. The commutators $[S_\epsilon, f_\eta^g]$ appear here rather than the products $S_\epsilon f_\eta^g$ because $\mathbf{S}$ operates both on f_η^g and on any ket which may ultimately appear to its right. For example,

$$\begin{aligned}S_z f_\eta^g \mid S'M'\rangle &= (S_z f_\eta^g) \mid S'M'\rangle + f_\eta^g S_z \mid S'M'\rangle \\ &= \eta f_\eta^g \mid S'M'\rangle + f_\eta^g S_z \mid S'M'\rangle\end{aligned} \tag{9.2}$$

an equation which illustrates the point.

If we now define the spin vectors $\mathbf{S}^1$ and $\mathbf{S}^2$, which are restricted so that they operate, respectively, only on f_η^g and only on $|S'M'\rangle$, we have

$$\begin{aligned}S_\epsilon^1 f_\eta^g \mid S'M'\rangle &= (S_\epsilon^1 f_\eta^g) \mid S'M'\rangle \\ &= [S_\epsilon^1, f_\eta^g] \mid S'M'\rangle\end{aligned} \tag{9.3}$$

in accord with Eqs. (9.1) and (9.2) and also

$$\begin{aligned}S_\epsilon^2 f_\eta^g \mid S'M'\rangle &= f_\eta^g S_\epsilon^2 \mid S'M'\rangle \\ \mathbf{S} &= \mathbf{S}^1 + \mathbf{S}^2\end{aligned} \tag{9.4}$$

$\mathbf{S}^1$ and $\mathbf{S}^2$ are now two commuting angular momenta whose sum is $\mathbf{S}$; hence, if we write

$$\langle S^1S^2M_1M_2 \mid S^1S^2S'M'\rangle$$

for the Wigner coefficients, we find that the kets

$$|\alpha S''M''\rangle = \sum_{\eta M'} \langle gS'\eta M' \mid gS'S''M''\rangle f_\eta^g \mid S'M'\rangle \tag{9.5}$$

are eigenstates of $\mathbf{S}^2$, S_z with eigenvalues $S''(S''+1)$, M'' and correctly connected in phase.

The inverse equation

$$f_\eta^g \mid S'M'\rangle = \sum_{S''M''} \langle gS'S''M'' \mid gS'\eta M'\rangle| \alpha S''M''\rangle \tag{9.6}$$

also holds. We now use the property

$$\langle SM \mid \alpha S''M''\rangle = A\delta_{SS''}\delta_{MM''}$$

for angular momentum eigenstates to deduce that

$$\langle SM \mid f^g_\eta \mid S'M'\rangle = A\langle gS'SM \mid gS'\eta M'\rangle \tag{9.7}$$

where A is independent of M, M', and η.

It is usual to define real $\overline{V}$ coefficients by the equation [(3), p. 50]

$$\langle gS'\eta M' \mid gS'SM\rangle = (-1)^{2S'+S-M}\sqrt{2S+1}\,\overline{V}\begin{pmatrix} g & S' & S \\ \eta & M' & -M\end{pmatrix} \tag{9.8}$$

We now follow Fano and Racah [(3), Eq. (14.4)] and define reduced matrix elements by the equation

$$\langle SM \mid f^g_\eta \mid S'M'\rangle = (-1)^{S-M}\langle S \,\|\, f^g \,\|\, S'\rangle\overline{V}\begin{pmatrix} S & S' & g \\ -M & M' & \eta\end{pmatrix} \tag{9.9}$$

in analogy with Eq. (2.34). Both Eq. (9.8) and Eq. (9.9) are merely definitions. It will appear in a moment that $\overline{V}$ has a simple behaviour under permutation of its columns. We will then be able to deduce from Eq. (9.7) that $\langle S \,\|\, f^g \,\|\, S'\rangle$ is independent of M, M', and η, as we obviously desire. Because of this independence, Eq. (9.9) may also be regarded as a theorem equivalent to Eq. (9.7). It is usually called the Wigner-Eckart theorem.

It follows from Eq. (9.8) that the $\overline{V}$ satisfy the orthonormality relations

$$\sum_{MM'} \overline{V}\begin{pmatrix} S & S' & S'' \\ M & M' & M''\end{pmatrix}\overline{V}\begin{pmatrix} S & S' & S''' \\ M & M' & M'''\end{pmatrix} = (2S''+1)^{-1}\delta_{S''S'''}\delta_{M''M'''}\delta(S, S', S'') \tag{9.10}$$

$$\sum_{S''M''} (2S''+1)\overline{V}\begin{pmatrix} S & S' & S'' \\ M & M' & M''\end{pmatrix}\overline{V}\begin{pmatrix} S & S' & S'' \\ M_1 & M_1' & M''\end{pmatrix} = \delta_{MM_1}\delta_{M'M_1'}$$

where $\delta(S, S', S'') = 1$ if $|S - S'| \le S'' \le S + S'$ and $S + S' + S''$ is integral, while $\delta(S, S', S'') = 0$ otherwise. If $S = 0$ in Eq. (9.8), we find

$$\overline{V}\begin{pmatrix} g & g' & 0 \\ \eta & -\eta' & 0\end{pmatrix} = (-1)^{g+\eta}(2g+1)^{-1/2}\delta_{gg'}\delta_{\eta\eta'} \tag{9.11}$$

Finally, $\overline{V}$ of Eq. (9.8) possesses various symmetry properties. First, it is invariant to even permutations of its columns and gets multiplied by $(-1)^{g+S+S'}$ by odd permutations. Secondly, we have

$$\overline{V}\begin{pmatrix} g & S & S' \\ \eta & M & M'\end{pmatrix} = (-1)^{g+S+S'}\overline{V}\begin{pmatrix} g & S & S' \\ -\eta & -M & -M'\end{pmatrix} \tag{9.12}$$

which is proved in (3), p. 51, and Wigner (4), p. 291, and which also follows from Griffith (26), example 7, p. 209. The behaviour of $\overline{V}$ under permutation of its columns is also established in Fano and Racah and in Wigner (same references).

In the present book we leave the strictly group-theoretic significance of the Racah-Wigner apparatus in the background. However, the reader should, at least, realize clearly that there exists a group U_2 of spin transformations [see Murnaghan (12), Chapter 10, Wigner (4), Chapter 15, or Griffith (26), pp. 169–180, for example] for which the set $|SM\rangle$ of $(2S + 1)$ spin eigenkets forms an irreducible representation of degree $(2S + 1)$. Once this fact has been appreciated, it becomes apparent that the analogy between the $\overline{V}$ (and derived coefficients) and the V (and W and X) goes very deep indeed.

9.2 $\overline{W}$ COEFFICIENTS

$\overline{W}$ coefficients are defined [(3), p. 54] as

$$\overline{W}\begin{pmatrix} a & b & c \\ d & e & f \end{pmatrix} = \sum_{\alpha\beta\gamma\delta\epsilon\phi} (-1)^{a-\alpha+b-\beta+c-\gamma+d-\delta+e-\epsilon+f-\phi} \overline{V}\begin{pmatrix} a & b & c \\ -\alpha & -\beta & -\gamma \end{pmatrix} \overline{V}\begin{pmatrix} a & e & f \\ \alpha & \epsilon & -\phi \end{pmatrix} \overline{V}\begin{pmatrix} b & f & d \\ \beta & \phi & -\delta \end{pmatrix} \overline{V}\begin{pmatrix} c & d & e \\ \gamma & \delta & -\epsilon \end{pmatrix} \tag{9.13}$$

The analogy with our previous definition [Eq. (4.1)] for the finite group W is apparent. $\overline{W}$, like W, is left invariant by any permutation of its columns or by turning any two of its columns upside down. It also satisfies an extra symmetry relation.* If we write

$$s_1 = a + d \qquad s_2 = b + e \qquad s_3 = c + f$$
$$d_1 = a - d \qquad d_2 = b - e \qquad d_3 = c - f$$

then the s_i and d_i determine the six quantities a, b, c, d, e, f. But $\overline{W}$ is invariant to separate permutation both of s_1, s_2, s_3 and of d_1, d_2, d_3.

If one of the symbols in Eq. (9.13) is zero, $\overline{W}$ may be simplified by using Eqs. (9.10) and (9.11). In particular,

$$\overline{W}\begin{pmatrix} 0 & b & c \\ d & e & f \end{pmatrix} = (-1)^{b+d+e}(2b + 1)^{-1/2}(2e + 1)^{-1/2}\delta_{bc}\delta_{ef}\delta(b, d, e) \tag{9.14}$$

In correspondence with Eqs. (4.10) and (4.11), we now have

$$\sum_{\delta\epsilon\phi} (-1)^{d-\delta+e-\epsilon+f-\phi} \overline{V}\begin{pmatrix} a & e & f \\ \alpha & \epsilon & -\phi \end{pmatrix} \overline{V}\begin{pmatrix} b & f & d \\ \beta & \phi & -\delta \end{pmatrix} \overline{V}\begin{pmatrix} c & d & e \\ \gamma & \delta & -\epsilon \end{pmatrix} = \overline{W}\begin{pmatrix} a & b & c \\ d & e & f \end{pmatrix} \overline{V}\begin{pmatrix} a & b & c \\ \alpha & \beta & \gamma \end{pmatrix} \tag{9.15}$$

* H. A. Jahn and K. M. Howell, *Proc. Camb. Phil. Soc.*, **55** (1959), 338; T. Regge, *Nuovo Cimento*, **11** (1959), 116.

$$\sum_{\phi} \overline{V}\begin{pmatrix} a & e & f \\ \alpha & \epsilon & -\phi \end{pmatrix} \overline{V}\begin{pmatrix} b & f & d \\ \beta & \phi & \delta \end{pmatrix}$$
$$= \sum_{c\gamma} (-1)^{c+f-\alpha-\delta}(2c+1)\overline{W}\begin{pmatrix} a & b & c \\ d & e & f \end{pmatrix} \overline{V}\begin{pmatrix} a & b & c \\ \alpha & \beta & \gamma \end{pmatrix} \overline{V}\begin{pmatrix} c & d & e \\ -\gamma & \delta & \epsilon \end{pmatrix} \tag{9.16}$$

There are also identities like Eqs. (4.13)–(4.16).

9.3 $\overline{X}$ COEFFICIENTS

To distinguish the next higher coefficient from our finite group coefficient X, we shall write it with a bar. It is defined by the equation [(3) Eq. (12.10)]

$$\overline{X}\begin{bmatrix} a & b & c \\ d & e & f \\ g & h & k \end{bmatrix} = \sum_{\alpha\beta\gamma\delta\epsilon\phi\eta\theta\kappa} \overline{V}\begin{pmatrix} a & b & c \\ \alpha & \beta & \gamma \end{pmatrix} \overline{V}\begin{pmatrix} d & e & f \\ \delta & \epsilon & \phi \end{pmatrix} \overline{V}\begin{pmatrix} g & h & k \\ \eta & \theta & \kappa \end{pmatrix} \overline{V}\begin{pmatrix} a & d & g \\ \alpha & \delta & \eta \end{pmatrix} \overline{V}\begin{pmatrix} b & e & h \\ \beta & \epsilon & \theta \end{pmatrix} \overline{V}\begin{pmatrix} c & f & k \\ \gamma & \phi & \kappa \end{pmatrix} \tag{9.17}$$

The quantity $\overline{X}$ of Eq. (9.17) satisfies

$$\overline{X} = \sum_{p} (-1)^{2p}(2p+1)\overline{W}\begin{pmatrix} p & e & a \\ b & c & h \end{pmatrix} \overline{W}\begin{pmatrix} p & c & h \\ k & g & f \end{pmatrix} \overline{W}\begin{pmatrix} p & g & f \\ d & e & a \end{pmatrix} \tag{9.18}$$

which is to be compared with Eq. (8.6) for X. When one symbol is zero, $\overline{X}$ reduces to a multiple of a single $\overline{W}$, and we have, for example,

$$X\begin{bmatrix} a & b & c \\ d & e & f \\ g & h & O \end{bmatrix} = (-1)^{b+d+f+h}(2c+1)^{-1/2}(2g+1)^{-1/2}\delta_{cf}\delta_{gh}\overline{W}\begin{pmatrix} a & b & c \\ e & d & g \end{pmatrix} \tag{9.19}$$

9.4 DOUBLE TENSOR OPERATORS

We possess definitions of reduced matrix elements with respect either to spin functions [Eq. (9.9)] or to space functions [Eq. (2.15)]. We now define matrix elements reduced with respect to both spin and space simultaneously. This definition is an entirely straightforward generalization of our previous work. The kets and bras are taken in an $S\Gamma M_s M_\Gamma$ scheme, and the operators are taken in a strictly analogous scheme. A typical double tensor operator is $X^{pb}_{\pi\beta}$, where, for fixed β and varying π, $X^{pb}_{\pi\beta}$ is an irreducible tensor operator with respect to spin which has $S = p$ and $M_s = \pi$. Similarly, for fixed π and varying β, $X^{pb}_{\pi\beta}$ is an irreducible tensor operator with respect to space for the irreducible representation b and component β.

A reduced matrix element of a double tensor operator is now defined by the equation

$$\langle SaM\alpha \mid X^{pb}_{\pi\beta} \mid S'a'M'\alpha' \rangle$$
$$= \langle Sa \,\|\, X^{pb} \,\|\, S'a' \rangle (-1)^{S-M} \overline{V} \begin{pmatrix} S & S' & p \\ -M & M' & \pi \end{pmatrix} V \begin{pmatrix} a & a' & b \\ \alpha & \alpha' & \beta \end{pmatrix} \tag{9.20}$$

and the reduced matrix element is independent of M, M', π, α, α' and β. We have written Eq. (9.20) assuming real component systems for our finite group representations. When we use complex systems, we replace

$$V \begin{pmatrix} a & a' & b \\ \alpha & \alpha' & \beta \end{pmatrix} \quad \text{with} \quad [-1]^{a+\alpha} V \begin{pmatrix} a & a' & b \\ -\alpha & \alpha' & \beta \end{pmatrix}$$

in accord with Eq. (2.34).

A particularly important kind of double tensor operator is a spin-independent operator. Then $p = \pi = 0$, and we shall often write such an operator as X^b_β for short. It can be reduced with respect either to space alone or to both spin and space; because of Eq. (9.11), the two kinds of reduced matrix element are related by the formula

$$\langle SaM \,\|\, X^b \,\|\, S'a'M' \rangle = (2S+1)^{-1/2} \delta_{SS'} \delta_{MM'} \langle Sa \,\|\, X^b \,\|\, S'a' \rangle \tag{9.21}$$

For a spin-independent operator x^b_β referring to a single electron, we shall often use the abbreviated notation

$$\langle a \,\|\, x^b \,\|\, a' \rangle \equiv \langle \tfrac{1}{2} a m_s \,\|\, x^b \,\|\, \tfrac{1}{2} a' m_s \rangle \tag{9.22}$$

which is consistent with our earlier use of this expression to denote a reduced matrix element for space functions alone, spin being completely ignored. Combining Eq. (9.21) with Eq. (9.22), we find the relation

$$\langle \tfrac{1}{2} a \,\|\, x^b \,\|\, \tfrac{1}{2} a' \rangle = \sqrt{2}\, \langle a \,\|\, x^b \,\|\, a' \rangle \tag{9.23}$$

We conclude this section by noting that the double tensor operators are defined with respect to two groups: the group U_2 for spin, mentioned in Section 9.1, and the group, G, say, for space. In their action on our operators, kets and bras in the $S\Gamma M_s M_\Gamma$ coupling scheme, the elements of U_2 commute with those of G. Hence, a double tensor operator for U_2 and G may equally be regarded as a single tensor operator for the direct product group $K = U_2 \times G$. The V and W coefficients for K are simply products of $\overline{V}$ and $\overline{W}$ for U_2 with V and W for G (see Section 3.2). Some of the formulae we shall derive shortly could be rendered more succinct in this way, but we do not do so, because it seems to make for greater clarity to treat spin and space as independently as possible.

9.5 SPIN-ORBIT COUPLING AS A DOUBLE TENSOR OPERATOR

The scalar product $\mathbf{s} \cdot \mathbf{l}$ may be regarded as a linear combination of components of one or more irreducible tensor operators which have $S = 1$ for the spin and whose space representations b are those which $\mathbf{l}$ spans under the spatial group G. If $G = O$ or T_d, there is just the one representation $b = T_1$. We shall consider only this situation explicitly, but we could easily generalize our treatment to deal with repeated representations. For the two groups just mentioned, the double tensor operator may be written $(sl)^{1T_1}_{M\beta}$ and satisfies

$$\mathbf{s} \cdot \mathbf{l} = \sum_{i=-1}^{1} (-1)^{1+i}(sl)^{1T_1}_{i-i} \tag{9.24}$$

when the complex tetragonal component system is adopted for T_1.

If we now allow the elements of O or T_d to operate on the components of $\mathbf{s}$ and $\mathbf{l}$ simultaneously, we can write $\mathbf{s} \cdot \mathbf{l}$ as an irreducible product, as in Eq. (5.6); i.e.,

$$\mathbf{s} \cdot \mathbf{l} = \sqrt{3}\,(s \times l)^{A_1} = \sum_{i=-1}^{1} (-1)^{1+i} s_i l_{-i} \tag{9.25}$$

Comparing Eq. (9.24) with Eq. (9.25) shows clearly the dual role of $\mathbf{s}$ as an irreducible spin tensor operator with $S = 1$ and as an irreducible tensor operator for the finite group belonging to the irreducible representation T_1. We have already used this duality in Chapter 6 to treat general two-electron systems, and it will appear shortly that spin-orbit coupling in n-electron systems having spin $S = 0$ or 1 can be treated in a very similar way.

Just as $S = 1$ corresponds to T_1, so $S = 0$ corresponds to A_1. In accord with this fact, our V coefficients in the complex tetragonal component system satisfy

$$V\begin{pmatrix} a & b & c \\ \alpha & \beta & \gamma \end{pmatrix} = \overline{V}\begin{pmatrix} a & b & c \\ \alpha & \beta & \gamma \end{pmatrix} \tag{9.26}$$

where each of a, b, c on the right-hand side is 0 or 1 and on the left-hand side is the corresponding one of A_1 or T_1. For example,

$$V\begin{pmatrix} A_1 & T_1 & T_1 \\ \alpha & \beta & \gamma \end{pmatrix} = \overline{V}\begin{pmatrix} 0 & 1 & 1 \\ \alpha & \beta & \gamma \end{pmatrix}$$

for all α, β, and γ. Furthermore, for A_1 or T_1, the equations

$$[-1]^a = (-1)^a, \qquad [-1]^\alpha = (-1)^\alpha \tag{9.27}$$

are true when, again, a is A_1 or T_1 on the left and 0 or 1, respectively, on the right. Hence, for A_1 and T_1 representations, the rule [Eq. (2.35)] for changing the signs of all components of V becomes identical with Eq. (9.12) for $\overline{V}$, as it should (note: $\overline{V} = 0$ unless $\alpha + \beta + \gamma = 0$; this is because the Wigner coefficients which correspond to them according to Eq. (9.8) are zero unless $-\gamma = \alpha + \beta$).

9.6 COUPLING OF SPIN WITH SPACE FOR OCTAHEDRAL (AND TETRAHEDRAL) SYMMETRY

In problems in which spin-orbit coupling is important, it is often convenient to use kets and bras which form bases for irreducible representations of the spinor symmetry group. In other words, we work in a scheme $Sht\tau$ rather than $ShM\theta$. Here t is a representation of the spinor group. If it occurs more than once in the term ${}^{2S+1}h$, we introduce an extra parameter to distinguish different occurrences. These parameters we shall denote by J (see Appendix E and (26), p. 242). Then we write the spin-orbit coupling energy as

$$\mathcal{H}_s = \sum_{\kappa=1}^{n} \mathbf{s}(\kappa) \cdot \mathbf{u}(\kappa) \tag{9.28}$$

In the $ShM\theta$ scheme, $\mathcal{H}_s$ would be regarded as a sum of components of double tensor operators, and we would find

$$\begin{aligned}\langle ShM\theta \mid \mathcal{H}_s \mid S'h'M'\theta'\rangle = \langle Sh \| \sum_{\kappa=1}^{n} su(\kappa) \| S'h'\rangle \sum_{i} (-1)^{i+1+S-M}[-1]^{h+\theta} \\ \overline{V}\begin{pmatrix} S & S' & 1 \\ -M & M' & i \end{pmatrix} V\begin{pmatrix} h & h' & T_1 \\ -\theta & \theta' & -i \end{pmatrix}\end{aligned} \tag{9.29}$$

In the $Sht\tau$ scheme, however, $\mathcal{H}_s$ is diagonal in t and τ; the diagonal elements are independent of τ and are given by

$$\begin{aligned}&\langle ShJt\tau \mid \mathcal{H}_s \mid S'h'J't\tau\rangle \\ &\quad = \sum_{MM'\theta\theta'} \langle ShJt\tau \mid ShM\theta\rangle\langle ShM\theta \mid \mathcal{H}_s \mid S'h'M'\theta'\rangle\langle S'h'M'\theta' \mid S'h'J't\tau\rangle \\ &\quad = \langle Sh \| \sum_{\kappa=1}^{n} su(\kappa) \| S'h'\rangle \Omega_{JJ'}\begin{pmatrix} S & S' & T_1 \\ h' & h & t \end{pmatrix}\end{aligned} \tag{9.30}$$

where

$$\begin{aligned}\Omega_{JJ'}\begin{pmatrix} S & S' & T_1 \\ h' & h & t \end{pmatrix} = \sum_{iMM'\theta\theta'} (-1)^{1+S-M'}[-1]^{h+\theta} \\ \overline{V}\begin{pmatrix} S & S' & 1 \\ -M & M' & i \end{pmatrix} V\begin{pmatrix} h & h' & T_1 \\ -\theta & \theta' & -i \end{pmatrix} \langle ShJt\tau \mid ShM\theta\rangle\langle S'h'M'\theta' \mid S'h'J't\tau\rangle\end{aligned} \tag{9.31}$$

using the fact that $\overline{V} = 0$ unless $i - M + M' = 0$. It follows easily from the symmetry properties of $\overline{V}$ and V, together with the rules [Eqs. (2.35) and (9.12)] for changing signs of components, that

$$\Omega_{J'J}\begin{pmatrix} S' & S & T_1 \\ h & h' & t \end{pmatrix} = (-1)^{S'+h'-S-h}\Omega_{JJ'}\begin{pmatrix} S & S' & T_1 \\ h' & h & t \end{pmatrix} \tag{9.32}$$

It is a simple matter of expansion to show that the Ω are proportional to recoupling coefficients according to the formula

$$\langle S1(S')h'J't\tau \mid S, 1h'(h)Jt\tau\rangle$$
$$= (-1)^{S-S'+h+h'}\lambda(h)^{1/2}(2S'+1)^{1/2}\Omega_{JJ'}\begin{pmatrix} S & S' & T_1 \\ h' & h & t \end{pmatrix} \quad (9.33)$$

and if S and S' are each either 0 or 1, we use the relation [Eq. (4.8)] between W and recoupling coefficients to deduce

$$\Omega\begin{pmatrix} S & S' & T_1 \\ h' & h & t \end{pmatrix} = (-1)^{S'+h+t+1}W\begin{pmatrix} S & S' & T_1 \\ h' & h & t \end{pmatrix} \quad (9.34)$$

It is possible to give further general formulae for various particular cases. The simplest of these, together with a considerable tabulation of values of Ω, are given in Appendix E. The possession of the values of Ω completes the passage from the $Sht\tau$ scheme to the $ShM\theta$ scheme, as far as a matrix element of spin-orbit coupling is concerned. The next task is to express the reduced matrix elements

$$\langle Sh \| \sum_{\kappa=1}^{n} su(\kappa) \| S'h'\rangle$$

for the n-electron system in terms of those for a one-electron system. This is accomplished with considerable generality in the next chapter.

9.7 MATRIX OF A SPIN-INDEPENDENT OPERATOR

Our treatment of spin-orbit coupling in the last section shows that even when a constant of proportionality between reduced matrix elements cannot be expressed in terms of W and $\overline{W}$, it may still be a multiple of a recoupling coefficient. The latter can then be calculated and its values tabulated.

Another example of such a factorization is furnished by a spin-independent operator, U^f_ϕ say. It was shown in Griffith (26), Eq. (8.32), [see also Koster (27)], that

$$\langle t\tau \mid U^f_\phi \mid t'\tau'\rangle = \sum_p a_p\langle ft'pt\tau \mid ft'\phi\tau'\rangle \quad (9.35)$$

where a_p is independent of τ, τ', and ϕ. t and t' are irreducible representations of the spinor group O^* (or T_d^*). If they are also representations of O or T_d, then Eq. (9.35) is equivalent to Eq. (2.15), and there is only one value of p. Otherwise, there may be two. It is convenient to rearrange Eq. (9.35) somewhat. Just as in Section 2.4, we can see that the coupling coefficients may be chosen to satisfy

$$\langle t'fpt\tau \mid t'f\tau'\phi\rangle = \epsilon\langle ft'pt\tau \mid ft'\phi\tau'\rangle \quad (9.36)$$

with ϵ independent of τ, τ', and ϕ. Then in place of Eq. (9.35) we can write

$$\langle t\tau \mid U^f_\phi \mid t'\tau'\rangle = \sum_p b_p \langle t'fpt\tau \mid t'f\tau'\phi\rangle \tag{9.37}$$

where $b_p = \epsilon a_p$. The problem now is to determine the b_p for matrix elements of U^f_ϕ in the $Sht\tau$ scheme in terms of the reduced matrix elements of U^f_ϕ in the $ShM\theta$ scheme. This is done by inverting Eq. (9.37) and then expanding. Thus:

$$\begin{aligned} b_p &= \sum_{\tau'\phi} \langle t'f\tau'\phi \mid t'fpt\tau\rangle \langle ShJt\tau \mid U^f_\phi \mid Sh'J't'\tau'\rangle \\ &= \lambda(h)^{-1/2} \langle Sh'(J't'), fpt \mid S, h'f(h)Jt\rangle \langle ShM \,\|\, U^f \,\|\, Sh'M\rangle \end{aligned} \tag{9.38}$$

In case S is 0 or 1, Eq. (9.38) achieves the simpler form

$$\begin{aligned} &\langle Sht \,\|\, U^f \,\|\, Sh't'\rangle \\ &= (-1)^{S+h'+t+f} \lambda(t)^{1/2} \lambda(t')^{1/2} W \begin{pmatrix} t & t' & f \\ h' & h & S \end{pmatrix} \langle ShM \,\|\, U^f \,\|\, Sh'M\rangle \end{aligned} \tag{9.39}$$

which agrees with Eq. (5.26).

CHAPTER 10

Matrices of One-Electron Operators

10.1 THE CONFIGURATIONS a^n

We now consider the general problem of determining the reduced matrix elements

$$\langle Sh \,||\, X^{pe} \,||\, S'h' \rangle$$

of a one-electron double tensor operator

$$X_{\pi\epsilon}^{pe} = \sum_{\kappa=1}^{n} x_{\pi\epsilon}^{pe}(\kappa) \tag{10.1}$$

between pairs of states of an n-electron configuration in terms of W, $\overline{W}$, X, $\overline{X}$, and the reduced matrix elements of $x_{\pi\epsilon}^{pe}$ for a one-electron system. This problem was considered originally by Tanabe and

Kamimura (21) following an analogous treatment by Racah (15) of atomic configurations. The present chapter is an exposition of Tanabe and Kamimura's work, using the formulation developed in this book; the formulae we derive become the same as theirs on appropriate change of formalism. However, although they contain nothing essentially new, the present formulae are considerably more general, because Tanabe and Kamimura expressed their results in terms of recoupling matrix elements and tabulated just enough of these to deal with the configurations $t_2^m e^n$ of the octahedral group. Because our results use W, $\overline{W}$, etc., they are immediately applicable to all problems for which these coefficients have been tabulated. W and X for all representations of the octahedral (hence, also T_d) and the dihedral groups D_3, D_4, D_5, D_∞ (also W, and implicitly X, for D_6) are given in Appendices C and D. Formulae for many special cases* and extensive tabulations of particular values† are available for $\overline{W}$ and $\overline{X}$.

First, we shall suppose that the states $|Sh\rangle$ and $|S'h'\rangle$ both belong to a configuration a^n. Then we simply generalize our treatment in Chapter 7 of spin-independent tensor operators to cover double tensor operators. The generalization is straightforward. We must extend the formulae given by Eqs. (5.25) and (5.26) to include spin; we obtain

$$\begin{aligned}\langle S_1aS_2bSc \,\|\, D^{pd} \,\|\, S_1'a'S_2'b'S'c'\rangle &= (-1)^{S'+S_1+S_2+p+S+a+b'+c'} \\ &\quad (2S+1)^{1/2}(2S'+1)^{1/2}\lambda(c)^{1/2}\lambda(c')^{1/2}\delta_{S_2S_2'}\delta_{bb'}\langle S_1a \,\|\, D^{pd} \,\|\, S_1'a'\rangle \\ &\quad \overline{W}\begin{pmatrix} S_1' & S_1 & p \\ S & S' & S_2 \end{pmatrix} W\begin{pmatrix} a' & a & d \\ c & c' & b \end{pmatrix}\end{aligned} \tag{10.2}$$

where S_1a, $S_1'a'$ refer to the first and S_2b, $S_2'b'$ to the second independent parts of the coupled states. $D^{pd}_{\pi\delta}$ is a double tensor operator for the first part of the system. Similarly, an operator for the second part satisfies

$$\begin{aligned}\langle S_1aS_2bSc \,\|\, E^{pe} \,\|\, S_1'a'S_2'b'S'c'\rangle &= (-1)^{S+S_1+S_2+p+e+a+b'+c} \\ &\quad (2S+1)^{1/2}(2S'+1)^{1/2}\lambda(c)^{1/2}\lambda(c')^{1/2}\delta_{S_1S_1'}\delta_{aa'}\langle S_2b \,\|\, E^{pe} \,\|\, S_2'b'\rangle \\ &\quad \overline{W}\begin{pmatrix} S_2' & S_2 & p \\ S & S' & S_1 \end{pmatrix} W\begin{pmatrix} b' & b & e \\ c & c' & a \end{pmatrix}\end{aligned} \tag{10.3}$$

The fractional parentage expression [Eq. (7.10)] for the states of a^n now achieves the simpler form

$$|a^nShM\theta\rangle = \sum_{S_1h_1} \langle a^{n-1}(S_1h_1), a|\} a^nSh\rangle|\, a^{n-1}(S_1h_1) \cdot a, ShM\theta\rangle \tag{10.4}$$

* References (2) and (3).

† K. M. Howell (1959), *Revised tables of 6j-symbols* (University of Southampton research report 59-1); Tables of 9j-symbols (idem, 59-2). M. Rotenberg, R. Bivins, N. Metropolis and J. K. Wooten, *The 3-j and 6-j symbols*. The Technology Press, Massachusetts Institute of Technology (1959).

The reduced matrix elements of $x^{pe}_{\pi\epsilon}$ for one electron are clearly zero unless $p = 0$ or 1. $p = 0$ is usually of interest only for spin-independent operators, which we have already discussed adequately for a^n in Section 7.2. The present approach would, of course, also give Eq. (7.14). The most interesting other operator is the spin-orbit coupling energy for the octahedral group, for which $p = 1$, $e = T_1$, and we readily deduce from Eqs. (10.3) and (10.4) that

$$\begin{aligned}\langle a^nSh \,\|\, \sum_{\kappa=1}^{n} su(\kappa) \,\|\, a^nS'h'\rangle &= n\langle a^nSh \,\|\, su(n) \,\|\, a^nS'h'\rangle \\ &= \langle \tfrac{1}{2}a \,\|\, su \,\|\, \tfrac{1}{2}a\rangle G_n(Sh, S'h')\end{aligned} \tag{10.5}$$

Clearly, $\langle \frac{1}{2}a \,\|\, su \,\|\, \frac{1}{2}a\rangle$ is zero unless $a = t_1$ or t_2. We shall write G_n^1 and G_n^2 to refer to these two cases. The formula for G_n^i is

$$\begin{aligned}G_n^i(Sh, S'h') = \sum_{S_1h_1} &\langle t_i^nSh\{|t_i^{n-1}(S_1h_1), t_i\rangle\langle t_i^{n-1}(S_1h_1), t_i|\}t_i^nS'h'\rangle \\ &(-1)^{S+S_1+\frac{1}{2}+h+h_1+T_i}n(2S+1)^{1/2}(2S'+1)^{1/2}\lambda(h)^{1/2}\lambda(h')^{1/2} \\ &W\begin{pmatrix}\frac{1}{2} & \frac{1}{2} & 1\\ S & S' & S_1\end{pmatrix} W\begin{pmatrix}T_i & T_i & T_1\\ h & h' & h_1\end{pmatrix}\end{aligned} \tag{10.6}$$

Clearly,

$$G_n^i(S'h', Sh) = (-1)^{S+h-S'-h'}G_n^i(Sh, S'h') \tag{10.7}$$

and

$$G_1^i(^2T_i, {}^2T_i) = 1, \qquad G_6^i(^1A_1, {}^1A_1) = 0$$

Another property of the G_n^i, which derives from the fact that the matrix of spin-orbit coupling for a^{6-n} is exactly the negative of that for a^n [at least for $n \neq 3$, see Griffith (26), Section 9.7], is

$$G_{6-n}^i(Sh, S'h') = -G_n^i(Sh, S'h'), \qquad n \neq 3$$

Armed with these results, we need only calculate G_n^i for $n = 2, 3$; those values are given in Table 10.1.

To proceed, we need also the reduced matrix elements $\langle \frac{1}{2}a \,\|\, su \,\|\, \frac{1}{2}a\rangle$. Actually, for any pair of one-electron functions a and a' for which we know the quantities $\langle a \,\|\, u \,\|\, a'\rangle$ it is very easy to deduce $\langle \frac{1}{2}a \,\|\, su \,\|\, \frac{1}{2}a'\rangle$. Consider the equation

$$\langle \tfrac{1}{2}am\alpha \mid s_0u_\beta \mid \tfrac{1}{2}a'm\alpha'\rangle = im\langle \tfrac{1}{2}am\alpha \mid u_\beta \mid \tfrac{1}{2}a'm\alpha'\rangle$$

which is true because $s_0 = is_z$, and express each side in terms of reduced matrix elements. The resulting equation simplifies to

$$\langle \tfrac{1}{2}a \,\|\, su \,\|\, \tfrac{1}{2}a'\rangle = \tfrac{1}{2}i\sqrt{6}\,\langle a \,\|\, u \,\|\, a'\rangle \tag{10.8}$$

and hence, as $\mathbf{u} = \zeta\mathbf{l}$, we can deduce the $\langle \frac{1}{2}a \,\|\, su \,\|\, \frac{1}{2}a'\rangle$ for p, d, or f

Table 10.1. Values of $G^i_n(Sb, S'b')$ for t_1^2, t_2^2, t_1^3 and t_2^3. The upper sign refers to t_1, and the lower to t_2, when there is a choice.

$n = 2$

Sb \ $S'b'$	$^3T_1$	$^1A_1$	^1E	$^1T_2$
$^3T_1$	1	$\frac{\sqrt{2}}{\sqrt{3}}$	$-\frac{1}{\sqrt{3}}$	$-\frac{1}{\sqrt{2}}$
$^1A_1$	$\frac{\sqrt{2}}{\sqrt{3}}$	·	·	·
^1E	$-\frac{1}{\sqrt{3}}$	·	·	·
$^1T_2$	$-\frac{1}{\sqrt{2}}$	·	·	·

$n = 3$

t_2^3	t_1^3	$^4A_2$ / $^4A_1$	^2E / ^2E	$^2T_2$ / $^2T_1$	$^2T_1$ / $^2T_2$
$^4A_2$	$^4A_1$	·	·	$\frac{2}{\sqrt{3}}$	·
^2E	^2E	·	·	$-\frac{\sqrt{2}}{\sqrt{3}}$	·
$^2T_2$	$^2T_1$	$\frac{2}{\sqrt{3}}$	$\frac{+\sqrt{2}}{-\sqrt{3}}$	·	1
$^2T_1$	$^2T_2$	·	·	-1	·

orbitals from the corresponding $\langle a \,\|\, l \,\|\, a' \rangle$, which were given in Table 2.5. The new reduced matrix elements are in Table 10.2.

As an illustrative example, let us evaluate the spin-orbit coupling energy of the A_1 level of the $^3T_1$ term of t_2^4 for f electrons. From Eqs. (9.30) and (10.5), we have

$$\langle f t_2^4\, {}^3T_1 A_1 \mid \mathcal{H}_s \mid f t_2^4\, {}^3T_1 A_1 \rangle$$

$$= \zeta \langle \tfrac{1}{2} t_2 \,\|\, sl \,\|\, \tfrac{1}{2} t_2 \rangle G_4^2({}^3T_1, {}^3T_1)\, \Omega \begin{pmatrix} 1 & 1 & T_1 \\ T_1 & T_1 & A_1 \end{pmatrix}$$

$$= -\tfrac{3}{2}\zeta \cdot -1 \cdot -W \begin{pmatrix} T_1 & T_1 & T_1 \\ T_1 & T_1 & A_1 \end{pmatrix} = -\tfrac{3}{2}\zeta \cdot -\tfrac{1}{3} = \tfrac{1}{2}\zeta$$

Table 10.2. Reduced matrix elements of the one-electron double tensor operator $s_\alpha u_\beta$ for p, d, and f electrons in units of the spin-orbit coupling parameter ζ.

p	T_1
T_1	-3

d	E	T_2
E	0	$-3\sqrt{2}$
T_2	$-3\sqrt{2}$	3

f	A_2	T_1	T_2
A_2	0	0	$-3\sqrt{2}$
T_1	0	$\frac{9}{2}$	$\frac{3}{2}\sqrt{15}$
T_2	$3\sqrt{2}$	$-\frac{3}{2}\sqrt{15}$	$-\frac{3}{2}$

10.2 MATRIX ELEMENTS BETWEEN a^n AND $a^{n-1}b$

In this case also, the matrix of a spin-independent operator was adequately discussed in Chapter 7, a discussion which is again readily generalized to deal with one-electron double tensor operators. For the spin-orbit coupling energy we obtain

$$\langle a^nSh \,||\, \sum_{\kappa=1}^{n} su(\kappa) \,||\, a^{n-1}(S_1h_1)bS'h'\rangle$$

$$= \langle \tfrac{1}{2}a \,||\, su \,||\, \tfrac{1}{2}b\rangle [n(2S+1)(2S'+1)\lambda(h)\lambda(h')]^{1/2} \quad (10.9)$$

$$(-1)^{S+S_1+\frac{1}{2}+h+h_1+b}\langle a^{n-1}(S_1h_1), a|\} a^nSh\rangle \overline{W}\begin{pmatrix} \frac{1}{2} & \frac{1}{2} & 1 \\ S & S' & S_1 \end{pmatrix} W\begin{pmatrix} a & b & T_1 \\ h' & h & h_1 \end{pmatrix}$$

and the relevant $\overline{W}$ are in Table 10.3.

Table 10.3. Values of $\overline{W}\begin{pmatrix} \frac{1}{2} & \frac{1}{2} & 1 \\ S & S' & S_1 \end{pmatrix}$ for various values of S, S', S_1.

S	0	$\frac{1}{2}$	$\frac{1}{2}$	$\frac{1}{2}$	1	1	$\frac{3}{2}$
S'	1	$\frac{1}{2}$	$\frac{1}{2}$	$\frac{3}{2}$	1	1	$\frac{3}{2}$
S_1	$\frac{1}{2}$	0	1	1	$\frac{1}{2}$	$\frac{3}{2}$	1
$\overline{W}$	$\frac{1}{\sqrt{6}}$	$\frac{1}{2}$	$\frac{1}{6}$	$-\frac{1}{3}$	$-\frac{1}{3}$	$-\frac{1}{6}$	$\frac{\sqrt{5}}{6\sqrt{2}}$

As an example, we evaluate $\mathcal{H}_s$ for d electrons between $t_2^6\,{}^1A_1$ and $t_2^5e\,{}^3T_1A_1$. Combining Eq. (9.30) with Eq. (9.34) gives

$$\langle t_2^6\,{}^1A_1 \mid \mathcal{H}_s \mid t_2^5e\,{}^3T_1A_1\rangle$$
$$= W\begin{pmatrix} A_1 & T_1 & T_1 \\ T_1 & A_1 & A_1 \end{pmatrix} \langle t_2^6\,{}^1A_1 \| \sum_{\kappa=1}^{6} su(\kappa) \| t_2^5e\,{}^3T_1\rangle$$
$$= \frac{1}{\sqrt{3}} \cdot -3\zeta\sqrt{2} \cdot 3\sqrt{6} \cdot -1 \cdot 1 \cdot \frac{1}{\sqrt{6}} \cdot -\tfrac{1}{3} = -\zeta\sqrt{6}$$

10.3 CONFIGURATIONALLY DIAGONAL MATRIX ELEMENTS FOR a^mb^n

These matrix elements are between two states, not necessarily the same, of a^mb^n. We need a preliminary proposition, which we prove in a rather general form so that it can also be used in another context in Section 10.4. Take a state $|X\rangle$ of an m-electron configuration and another $|Y\rangle$ of an n-electron configuration having no one-electron orbitals in common with the first. Then the state

$$|Z\rangle = [m!n!(m+n)!]^{-1/2} \sum_{\mu} (-1)^{\mu} P_{\mu} \mid X\rangle \cdot |Y\rangle \tag{10.10}$$

is a properly antisymmetrized and normalized state of the composite $(m+n)$-electron configuration. Let

$$|Z'\rangle = [m!n!(m+n)!]^{-1/2} \sum_{\nu} (-1)^{\nu} P_{\nu} \mid X'\rangle \cdot |Y'\rangle \tag{10.11}$$

be another such state, but do not suppose that the new $(m+n)$-electron configuration is necessarily the same. Suppose merely that the two configurations of which $|X\rangle$ and $|Y'\rangle$ are constituent states have no orbitals in common and suppose a similar relationship for $|X'\rangle$ and $|Y\rangle$. If $F = \Sigma f(\kappa)$ is a one-electron operator, then

$$\langle Z' \mid F \mid Z\rangle = [m!n!(m+n)!]^{-1/2} \sum_{\mu} \langle Z' \mid (-1)^{\mu} P_{\mu} F \mid X\rangle \cdot |Y\rangle$$
$$= \left(\frac{(m+n)!}{m!n!}\right)^{1/2} \langle Z' \mid F \mid X\rangle \cdot |Y\rangle \tag{10.12}$$

where we used the antisymmetry of $\langle Z'|$. Next, note that no part of $\langle Z'|$ for which the first m electrons are not in $\langle X'|$ can give any contribution to the matrix element, because $|X\rangle$, $|Y\rangle$ have, respectively, no orbitals in common with $|Y'\rangle$, $|X'\rangle$. So only the $m!n!$ terms in the sum [Eq. (10.11)] which have the first m electrons in $|X'\rangle$ need be considered. Hence, finally,

$$\langle Z' \mid F \mid Z\rangle = \langle X' \mid F \mid X\rangle + \langle Y' \mid F \mid Y\rangle \tag{10.13}$$

Formulae for the matrix elements of a double tensor operator within a^mb^n now follow at once from Eqs. (10.2) and (10.3). For the spin-orbit coupling energy, we find

$$\begin{aligned}
&\langle a^m(S_1h_1)b^n(S_2h_2)Sh \,\|\, \textstyle\sum su(\kappa) \,\|\, a^m(S_1'h_1')b^n(S_2'h_2')S'h'\rangle \\
&\quad = (-1)^{S_1+S_2+h_1+h_2'}(2S+1)^{1/2}(2S'+1)^{1/2}\lambda(h)^{1/2}\lambda(h')^{1/2} \\
&\quad \Bigg[(-1)^{S'+h'}\delta_{S_2S_2'}\delta_{h_2h_2'}\overline{W}\begin{pmatrix} S_1' & S_1 & 1 \\ S & S' & S_2\end{pmatrix} W\begin{pmatrix} h_1' & h_1 & T_1 \\ h & h' & h_2\end{pmatrix} \\
&\qquad\qquad G_n(S_1h_1, S_1'h_1')\langle \tfrac{1}{2}a \,\|\, su \,\|\, \tfrac{1}{2}a\rangle \\
&\quad + (-1)^{S+h}\delta_{S_1S_1'}\delta_{h_1h_1'}\overline{W}\begin{pmatrix} S_2' & S_2 & 1 \\ S & S' & S_1\end{pmatrix} W\begin{pmatrix} h_2' & h_2 & T_1 \\ h & h' & h_1\end{pmatrix} \\
&\qquad\qquad G_n(S_2h_2, S_2'h_2')\langle \tfrac{1}{2}b \,\|\, su \,\|\, \tfrac{1}{2}b\rangle\Bigg]
\end{aligned} \tag{10.14}$$

A particular case of special importance for the theory of transition-metal ions occurs when $a = t_2$, $b = e$ and then the second term of Eq. (10.14) is identically zero. As an example, let us evaluate the diagonal element of $\mathcal{H}_s$ for $t_2^5e\ {}^3T_1T_1$. We use Eq. (9.30) and then Eq. (10.14) and Table 10.3 to give

$$\begin{aligned}
\langle t_2^5e\ {}^3T_1T_1 \mid \mathcal{H}_s \mid t_2^5e\ {}^3T_1T_1\rangle &= \tfrac{1}{6}\langle t_2^5e\ {}^3T_1 \,\|\, \textstyle\sum su(\kappa) \,\|\, t_2^5e\ {}^3T_1\rangle \\
&= \tfrac{1}{12}\langle \tfrac{1}{2}t_2 \,\|\, su \,\|\, \tfrac{1}{2}t_2\rangle = \tfrac{1}{4}\zeta
\end{aligned}$$

In general, we shall need $\overline{W}$ which are not given in Table 10.3. Further ones are not given in this book, because extensive tables are available elsewhere (see page 86). A similar remark applies to the $\overline{X}$ coefficients. In the literature the $\overline{W}$ and $\overline{X}$ are called $\overline{W}$ or $6j$ symbols and X or $9j$ symbols, respectively.

For a spin-independent operator U_ϕ^f, all matrix elements are zero which do not satisfy $S = S'$, $M = M'$, $S_1 = S_1'$, and $S_2 = S_2'$. When these conditions are satisfied,

$$\begin{aligned}
&\langle a^m(S_1h_1)b^n(S_2h_2)ShM \,\|\, U^f \,\|\, a^m(S_1h_1')b^n(S_2h_2')Sh'M\rangle \\
&= (-1)^{f+h_1+h_2'}\lambda(h)^{1/2}\lambda(h')^{1/2}\Bigg[(-1)^{h'}\delta_{h_2h_2'}W\begin{pmatrix} h_1' & h_1 & f \\ h & h' & h_2\end{pmatrix} g^m_{h_1h_1'}(a,f)\langle a \,\|\, u^f \,\|\, a\rangle \\
&+ (-1)^{h}\delta_{h_1h_1'}W\begin{pmatrix} h_2' & h_2 & f \\ h & h' & h_1\end{pmatrix} g^n_{h_2h_2'}(b,f)\langle b \,\|\, u^f \,\|\, b\rangle\Bigg]
\end{aligned} \tag{10.15}$$

The quantities $g^n_{hh'}(a, f)$ were defined and equations and values given in Chapter 7.

10.4 MATRIX ELEMENTS BETWEEN $a^{m-1}b^n$ AND a^mb^{n-1}

Here again, we make a preliminary investigation of what may be called the permutational part of the simplification of a matrix element between a typical bra $\langle Z'|$ of $a^{m-1}b^n$

and a typical ket $|Z\rangle$ of $a^m b^{n-1}$ and let $|X\rangle$, $|X'\rangle$, $|Y\rangle$, $|Y'\rangle$ belong to a^m, a^{m-1}, b^{n-1}, b^n, respectively. Then, at the stage [Eq. (10.12)] of the reduction, observe that only those parts of F which connect an orbital a of $|X\rangle$ with an orbital b of $|Y'\rangle$ can give a non-zero contribution. So we have the relation

$$F = \sum_{\kappa=1}^{m+n} f(\kappa) \doteqdot \sum_{\kappa=1}^{m} f(\kappa) \doteqdot mf(m)$$

where in the last replacement we utilize the antisymmetry of $\langle Z'|$ and $|X\rangle$. After that, we expand the state $|X\rangle$ by a fractional parentage procedure, getting the mth electron always on the right. Then we simplify and finally expand $\langle Y'|$, getting the mth electron always at the left. Schematically:

$$\begin{aligned}\langle a^{m-1}b^n \mid F \mid a^m b^{n-1}\rangle &= \left[\frac{m(m+n)!}{(m-1)!(n-1)!}\right]^{1/2} \langle a^{m-1}b^n \mid f(m) \mid a^m\rangle \cdot |b^{n-1}\rangle \\ &= \sqrt{mn}\,\langle a^{m-1}, a|\}a^m\rangle\langle b^n \mid f(m) \mid a\rangle \cdot |b^{n-1}\rangle \qquad (10.16) \\ &= \sqrt{mn}\,\langle a^{m-1}, a|\}a^m\rangle\langle b^n\{|b, b^{n-1}\rangle\langle b \mid f \mid a\rangle\end{aligned}$$

The formula for a typical matrix element is now obtained by straightforward but complicated expansion. First, we invert Eq. (9.20) and then perform all the expansions needed to reach Eq. (10.16). Finally, we express the coupling coefficients in terms of V and $\overline{V}$, and $\langle b \mid f \mid a\rangle$ in terms of its reduced matrix element. The sum thus obtained contains products of six V and six $\overline{V}$, which are easily recognized to be X and $\overline{X}$ coefficients, respectively. The final result is

$$\begin{aligned}&\langle a^{m-1}(S_1h_1)b^n(S_2h_2)Sh \,\|\, \textstyle\sum su(\kappa) \,\|\, a^m(S_1'h_1')b^{n-1}(S_2'h_2')S'h'\rangle \\ &= (-1)^{S_1-S_1'-S_2+S_2'+h_1+h_1'+h_2+h_2'} \\ &\quad [mn(2S+1)(2S'+1)(2S_1'+1)(2S_2+1)\lambda(h)\lambda(h')\lambda(h_1')\lambda(h_2)]^{1/2} \\ &\quad \langle a^{m-1}(S_1h_1), a|\}a^m S_1'h_1'\rangle\langle b^n S_2h_2\{|b, b^{n-1}(S_2'h_2')\rangle \qquad (10.17) \\ &\quad \overline{X}\begin{bmatrix} S_1 & S_2 & S \\ S_1' & S_2' & S' \\ \frac{1}{2} & \frac{1}{2} & 1 \end{bmatrix} X \begin{bmatrix} h_1 & h_2 & h \\ h_1' & h_2' & h' \\ a & b & T_1 \end{bmatrix} \langle \tfrac{1}{2}b \,\|\, su \,\|\, \tfrac{1}{2}a\rangle\end{aligned}$$

for the spin-orbit coupling. It is useful to have also the related and slightly simpler formula

$$\begin{aligned}&\langle a^m(S_1h_1)b^{n-1}(S_2h_2)Sh \,\|\, \textstyle\sum su(\kappa)\|\, a^{m-1}(S_1'h_1')b^n(S_2'h_2')S'h'\rangle \\ &\qquad = [mn(2S+1)(2S'+1)(2S_1+1)(2S_2'+1)\lambda(h)\lambda(h')\lambda(h_1)\lambda(h_2')]^{1/2} \\ &\qquad \langle a^{m-1}(S_1'h_1'), a|\}a^m S_1h_1\rangle\langle b^n S_2'h_2'\{|b, b^{n-1}(S_2h_2)\rangle \qquad (10.18)\end{aligned}$$

$$\overline{X}\begin{bmatrix} S_1 & S_2 & S \\ S_1' & S_2' & S' \\ \frac{1}{2} & \frac{1}{2} & 1 \end{bmatrix} X \begin{bmatrix} h_1 & h_2 & h \\ h_1' & h_2' & h' \\ a & b & T_1 \end{bmatrix} \langle \tfrac{1}{2}a \,||\, su \,||\, \tfrac{1}{2}b\rangle$$

For a spin-independent operator U_ϕ^f, we obtain

$$\begin{aligned} &\langle a^{m-1}(S_1h_1)b^n(S_2h_2)ShM \,||\, U^f \,||\, a^m(S_1'h_1')b^{n-1}(S_2'h_2')Sh'M\rangle \\ &\quad = (-1)^{S_1+S_2'+S+\frac{1}{2}+h_1+h_1'+h_2+h_2'+f} \\ &\qquad [mn(2S_1'+1)(2S_2+1)\lambda(h)\lambda(h')\lambda(h_1')\lambda(h_2)]^{1/2} \\ &\qquad \langle a^{m-1}(S_1h_1), a|\} a^m S_1'h_1'\rangle\langle b^n S_2h_2\{|b, b^{n-1}(S_2'h_2')\rangle \qquad (10.19) \\ &\qquad \overline{W}\begin{pmatrix} S_1 & S_2 & S \\ S_4 & S_3 & \frac{1}{2} \end{pmatrix} X \begin{bmatrix} h_1 & h_2 & h \\ h_1' & h_2' & h' \\ a & b & f \end{bmatrix} \langle b \,||\, u^f \,||\, a\rangle \end{aligned}$$

and

$$\begin{aligned} &\langle a^m(S_1h_1)b^{n-1}(S_2h_2)ShM \,||\, U^f \,||\, a^{m-1}(S_1'h_1')b^n(S_2'h_2')Sh'M\rangle \\ &\quad = (-1)^{S_1'+S_2+S+\frac{1}{2}} [mn(2S_1+1)(2S_2'+1)\lambda(h)\lambda(h')\lambda(h_1)\lambda(h_2')]^{1/2} \\ &\qquad \langle a^{m-1}(S_1'h_1'), a|\} a^m S_1h_1\rangle\langle b^n S_2'h_2'\{|b, b^{n-1}(S_2h_2)\rangle \qquad (10.20) \\ &\qquad \overline{W}\begin{pmatrix} S_1 & S_2 & S \\ S_4 & S_3 & \frac{1}{2} \end{pmatrix} X \begin{bmatrix} h_1 & h_2 & h \\ h_1' & h_2' & h' \\ a & b & f \end{bmatrix} \langle a \,||\, u^f \,||\, b\rangle \end{aligned}$$

The formulae we have derived in Sections (10.3) and (10.4) are also the kernel of the reduction of matrix elements of a one-electron operator between states belonging to configurations having three or more different constituent representations. Such matrix elements are zero if the configurations differ in more than one constituent orbital. Hence, between $a^m b^n c^o \ldots$ and $a^{m'} b^{n'} c^{o'} \ldots$, at most two of the inequalities $m \neq m'$, $n \neq n'$, $o \neq o'$, ... can hold; i.e., at least one equality must hold. Suppose the equality to hold for m and m' and that, further, in the coupled states of both configurations, a^m is coupled last to the rest of the configuration. We have already established Eq. (10.13) in a sufficiently general form for us then to be able to write equations analogous to Eqs. (10.14) and (10.15). In these, b^n is replaced with $b^n c^o \ldots$ in the bra and with $b^{n'} c^{o'}$ in the ket. The right-hand sides of the equation can no longer be simplified by the formulae given by Eqs. (10.5) and (7.14). So, if $b^n c^o \ldots$ and $b^{n'} c^{o'} \ldots$ are coupled to S_2h_2 and $S_2'h_2'$ respectively, then the second of each of G_n and g^n are replaced according to the rules

$$G_n(S_2h_2, S_2'h_2')\langle \tfrac{1}{2}b \,||\, su \,||\, \tfrac{1}{2}b\rangle \longrightarrow \langle S_2h_2 \,||\, \textstyle\sum su(\kappa) \,||\, S_2'h_2'\rangle$$

$$g^n_{h_2h_2'}(b, f)\langle b \,||\, u^f \,||\, b\rangle \longrightarrow \langle S_2h_2M_2 \,||\, U^f \,||\, S_2h_2'M_2\rangle$$

In other cases some recoupling and reordering of $a, b, c, \ldots$ and $a', b', c', \ldots$ may be necessary, but this can always be accomplished by using W, $\overline{W}$, or higher coefficients (cf. Section 4.3 and Ref. 2–5). Hence, we can always eliminate one kind of one-electron orbital, and, therefore, we can evaluate any matrix element by eliminating each kind of orbital in turn until we get down to matrix elements involving only two kinds of orbital, for all of which we already have general formulae.

CHAPTER 11

Some Special Topics

11.1 A TRIGONAL LIGAND FIELD FOR *d* ELECTRONS

Here we suppose that we have a predominantly octahedral field with a trigonal component along the (111) axis. Classifying it in terms of the octahedral group, the trigonal field will be a one-electron operator which is a sum of those components of irreducible tensor operators for the octahedral group which possess full trigonal symmetry about the (111) axis. There are two distinct kinds of components satisfying this condition; one has A_1 symmetry, and the other is the τ component of the T_2 representation in the trigonal component system. The first is not of much interest, since it is most conveniently incorporated with the rest of the octahedral field. The second can have non-zero matrix elements between t_2 and e or within t_2, but not within e. It could be written $u_\tau^{T_2}$ when it refers to one electron, but we shall write it u_τ for short, and $U_\tau = \Sigma\, u_\tau(\kappa)$ for an n-electron system.

u_τ is completely characterized within the t_2 and e orbitals by specifying the two reduced matrix elements $\langle t_2 \| u \| t_2\rangle$ and $\langle t_2 \| u \| e\rangle$. It splits the t_2 orbitals into a singlet lying at

$$\langle t_2\tau \mid u_\tau \mid t_2\tau\rangle = \langle t_2 \| u \| t_2\rangle V\begin{pmatrix} T_2 & T_2 & T_2 \\ \tau & \tau & \tau \end{pmatrix}$$
$$= -\frac{\sqrt{2}}{3}\langle t_2 \| u \| t_2\rangle$$

and a doublet at

$$\langle t_2\sigma \mid u_\tau \mid t_2\sigma\rangle = \langle t_2 \| u \| t_2\rangle V\begin{pmatrix} T_2 & T_2 & T_2 \\ \sigma & \sigma & \tau \end{pmatrix} = \frac{1}{3\sqrt{2}}\langle t_2 \| u \| t_2\rangle$$

Hence, if the singlet lies lower than the doublet by an amount δ, we find

$$\langle t_2 \| u \| t_2\rangle = \delta\sqrt{2} \qquad (11.1)$$

U_τ splits only T_1 and T_2 terms, and, just as above, the singlet lies below the doublet by an amount

$$D = \frac{1}{\sqrt{2}}\langle {}^{2S+1}T_iM \| U \| {}^{2S+1}T_iM\rangle$$

In the strong-field coupling scheme we then deduce from Eqs. (10.15) and (11.1) that for a term ${}^{2S+1}T_i$ from the configuration $t_2^m(S_1h_1)e^n(S_2h_2)$:

$$D = 3\delta(-1)^{h_1+h_2+T_i}\, W\begin{pmatrix} h_1 & h_1 & T_2 \\ T_i & T_i & h_2 \end{pmatrix} g^m_{h_1h_1}(T_2, T_2) \qquad (11.2)$$

Among the ground terms of d^n configurations we get $t_2^2(^3T_1)$ for d^2, $t_2^4(^3T_1)e^2(^3A_2)^5T_2$ for d^6, and $t_2^5(^2T_2)e^2(^3A_2)^4T_1$ for d^7. It is a mere matter of substitution in Eq. (11.2) to find $-\delta$, δ, $-\delta$, respectively, for D. Naturally, any other part of the matrix of U_τ is equally easily deduced from the formulae of Chapter 10.

REFERENCES

Jarrett, H. S., *J. Chem. Phys.*, **27** (1957), 1298.
Piper, T. S. and R. L. Carlin, *J. Chem. Phys.*, **33** (1960), 1208.
Pryce, M. H. L. and W. A. Runciman, *Disc. Farad. Soc.*, **26** (1958), 34.
Sugano, S., Y. Tanabe, and I. Tsujikawa, *J. Phys. Soc. Japan*, **13** (1958), 880.
Van Vleck, J. H., *Disc. Farad. Soc.*, **26** (1958), 96.

11.2 INTENSITY CALCULATIONS FOR CENTROSYMMETRIC OCTAHEDRAL SYSTEMS

We consider here the calculation of the matrix elements of the electric dipole moment p_δ, transforming as T_1, for a transition between a ground state orbital $|a\alpha\rangle$ and an excited orbital $|b\beta\rangle$ having the same parity. It is assumed to be accompanied by an absorption of one vibrational quantum, belonging to the component γ of a representation c, into an initially unexcited vibrational state. This transition is mediated by an orbital $|e\epsilon\rangle$ of opposite parity to $|a\alpha\rangle$ and $|b\beta\rangle$ under the influence of a nuclear-electronic interaction Hamiltonian

$$\mathcal{H}_v = \sum_\gamma q_\gamma^c Q_\gamma^c \tag{11.3}$$

where q refers to electrons and Q to nuclear normal coordinates. Then the actual transition is between the modified states

$$\begin{aligned} |X\rangle &= |a\alpha, 0\rangle - E^{-1} \sum_\epsilon \langle e\epsilon, 1 \mid \mathcal{H}_v \mid a\alpha, 0\rangle| e\epsilon, 1\rangle \\ |Y\rangle &= |b\beta, 1\rangle - E_1^{-1} \sum_\epsilon \langle e\epsilon, 0 \mid \mathcal{H}_v \mid b\beta, 1\rangle| e\epsilon, 0\rangle \end{aligned} \tag{11.4}$$

where E, E_1 are the energies of e above a, b, respectively, and 1 represents one quantum of the vibration. Then

$$\begin{aligned} \langle X|p_\delta|Y\rangle &= -\left(\frac{\hbar}{2\nu}\right)^{1/2} \sum_\epsilon [E^{-1}\langle a\alpha \mid q_\gamma^c \mid e\epsilon\rangle\langle e\epsilon \mid p_\delta \mid b\beta\rangle + E_1^{-1}\langle a\alpha \mid p_\delta \mid e\epsilon\rangle\langle e\epsilon \mid q_\gamma^c \mid b\beta\rangle] \\ &= \sum_\epsilon \left[R_1 V \begin{pmatrix} a & e & c \\ \alpha & \epsilon & \gamma \end{pmatrix} V \begin{pmatrix} e & b & T_1 \\ \epsilon & \beta & \delta \end{pmatrix} + R_2 V \begin{pmatrix} a & e & T_1 \\ \alpha & \epsilon & \delta \end{pmatrix} V \begin{pmatrix} e & b & c \\ \epsilon & \beta & \gamma \end{pmatrix}\right] \end{aligned} \tag{11.5}$$

where

$$R_1 = -\left(\frac{\hbar}{2\nu}\right)^{1/2} E^{-1}\langle a \,\|\, q^c \,\|\, e\rangle\langle e \,\|\, p \,\|\, b\rangle$$

$$R_2 = -\left(\frac{\hbar}{2\nu}\right)^{1/2} E_1^{-1}\langle a \,\|\, p \,\|\, e\rangle\langle e \,\|\, q^c \,\|\, b\rangle$$

Often we are interested in the total transition probability from a particular state $|a\alpha\rangle$ to all components β, γ and using unpolarized light. We shall then evaluate

$$\begin{aligned} \sum_{\beta\gamma\delta} |\langle X \mid p_\delta \mid Y\rangle|^2 &= \lambda(a)^{-1}\lambda(e)^{-1}R_1^2 \\ &\quad + 2(-1)^{c+1}\lambda(a)^{-1}R_1R_2W \begin{pmatrix} a & e & T_1 \\ b & e & c \end{pmatrix} + \lambda(a)^{-1}\lambda(e)^{-1}R_2^2 \end{aligned} \tag{11.6}$$

Note that even in this one-electron case we get a W coefficient in the cross term R_1R_2, but not in the "direct" terms R_1^2 and R_2^2. For n-electron transitions, the situation is more complicated and the results cannot usually be expressed in terms of single W or X coefficients. N. K. Hamer has investigated the theory of transitions between $a^m b^{n-1}$ and $a^{m-1} b^n$ and finds that products of twelve V coefficients can occur (to be published).

It is interesting to mention, however, that there are two distinct kinds of coefficient involving a sum of products of eight V coefficients. We may call them Y and Y' and they may be related, respectively, to a simple endless strip and a Mobius (or twisted) strip. Since these matters are beyond the scope of this book, the reader is referred to the references given below for a description of the corresponding higher coefficients for spherical symmetry.

REFERENCES FOR INTENSITY THEORY

Ballhausen, C. J. and A. D. Liehr, *Mol. Phys.*, **2** (1959), 123.
Englman, R., *Mol. Phys.*, **3** (1960), 48.
Griffith, J. S., *Mol. Phys.*, **3** (1960), 477.
Koide, S., *Phil. Mag.*, **4** (1959), 243.
Koide, S. and M. H. L. Pryce, *Phil. Mag.*, **3** (1959), 607.
Liehr, A. D. and C. J. Ballhausen, *Phys. Rev.*, **106** (1957), 1161.
Pappalardo, R., *J. Chem. Phys.*, **31** (1959), 1050.
Sugano, S., *Prog. Theor. Phys.*, Supplement No. 14 (1960), 66.
Tanabe, Y., (1960), *idem*, p. 17.

REFERENCES TO HIGHER COEFFICIENTS

Elliott, J. P. and B. H. Flowers, *Proc. Roy. Soc. A*, **229** (1955), 545.
Jahn, H. A. and J. Hope, *Phys. Rev.*, **93** (1954), 318.
Jucys, A. P., I. B. Levinson, and V. V. Vanagas, ref. 5 in Appendix F.
Ord-Smith, R. J., *Phys. Rev.*, **94** (1954), 1227.

11.3 SECOND-ORDER SPIN-ORBIT COUPLING IN FREE ATOMS

We now consider briefly a particularly interesting example from the theory for spherical symmetry. It parallels that of Section 5.4 very closely, but we rely more heavily on Fano and Racah's book (3) than previously.

In the *SLJM* coupling scheme, the spin-orbit coupling energy, $\mathcal{H}_s$, is diagonal in J and M and independent of M. It is a sum of irreducible products of degree 0 of a spin tensor operator of degree 1 with an orbital operator of degree 1. The second-order modification to the energy of the SLJ level by a level $S'L'J$ of another term lying at energy E higher is

$$E(J) = -E^{-1}|\langle SLJM \mid \mathcal{H}_s \mid S'L'JM\rangle|^2$$

which by formula (15.6) in Fano and Racah becomes

$$E(J) = R\overline{W}\begin{pmatrix} L' & S' & J \\ S & L & 1 \end{pmatrix}^2 \tag{11.7}$$

with R independent of J. Then, using the Biedenharn identity [(3), p. 159)], we deduce

$$E(J) = \sum_{g=0}^{2} c(g)\overline{W}\begin{pmatrix} g & S & S \\ J & L & L \end{pmatrix} \tag{11.8}$$

with $c(g)$ independent of J.

On the other hand, consider the quantity

$$Q = \sum_g a(g)[S^{(g)} \times L^{(g)}]^0 \tag{11.9}$$

which is a sum of irreducible products of degree 0 and where $S^{(g)}$ and $L^{(g)}$ are irreducible products of degree g of g spin vectors $\mathbf{S}$ and g orbital vectors $\mathbf{L}$, respectively. Again using formula (15.6) in Fano and Racah, we find that the matrix elements of Q within the term SL are diagonal in J and M and independent of M, with diagonal elements

$$\langle SLJM \mid Q \mid SLJM\rangle = \sum_g a'(g)\overline{W}\begin{pmatrix} g & S & S \\ J & L & L \end{pmatrix} \tag{11.10}$$

Hence, if we take Q as an operator equivalent within the term SL to the second-order correction to the energy caused by the term $S'L'$, we need only take the values $g = 0, 1, 2$ in the sum. The part $g = 0$ is a uniform shift of all states of the term, and the part $g = 1$ has the form $\lambda\mathbf{S} \cdot \mathbf{L}$ and hence the same form as the first-order effect of spin-orbit coupling within the term. The part $g = 2$ is proportional to $(\mathbf{S} \cdot \mathbf{L})^2 + \frac{1}{2}\mathbf{S} \cdot \mathbf{L} - \frac{1}{3}L(L + 1)S(S + 1)$, which has the same form as the first-order effect of the magnetic spin-spin coupling energy within the SL term. It is found empirically that the energies of levels of terms are usually well represented by the operator equivalent [Eq. (11.9)] with $a(g) = 0$ for $g > 2$.

REFERENCES

Araki, G., *Prog. Theor. Phys.*, **3** (1948), 152, 262.
Judd, B. R., *Proc. Phys. Soc. A*, **69** (1956), 157.
Pryce, M. H. L., *Phys. Rev.*, **80** (1950), 1107.
Trees, R. E., *Phys. Rev.*, **82** (1951), 683.

11.4 AROMATIC HYDROCARBONS

The most obvious potential applications of the irreducible tensor method to π-electron systems are given by cyclo-butadiene and benzene. Here the molecular orbitals form bases for irreducible representations of D_{4h} and D_{6h}, respectively. The same approach may be applied to heterocyclics of the first-row elements by regarding them as perturbations of the parent symmetrical hydrocarbon. Thus pyridine is derived from the isoelectronic benzene by moving one proton into a carbon nucleus, so that one CH group is turned into a nitrogen atom. The resulting perturbation is a one-electron operator which may be expanded as a sum of components of irreducible tensors relative to D_{6h}. Hence, general formulae for its matrix elements could be derived according to the methods of the present book.

Before we discuss larger aromatic systems, it is well to emphasize two things. One is the extreme power and ease of use of the irreducible tensor method. The formulae are superficially complicated, but they are in reality very straightforward to derive and manipulate. The second is the increasing ease of solving large matrix equations with computers. Consequently, a theoretical scheme in which the matrices of quantities of interest are easily and mechanically obtainable may, in the future, be of as much value as one in which they are not, even if the basic functions used in the scheme give a much better zero-order approximation to eigenstates of the energy in the second case than in the first.

Appreciating these facts, we may consider a scheme for naphthalene in which the basic functions are built up from molecular orbitals calculated for the periphery only.* In order to do so, we neglect, at first, the resonance integral (and overlap integral) across the central bond. The π-system then has symmetry group D_{10h}. The correction caused by the actual presence of the central bond is a one-electron operator which is a sum of components of irreducible tensor operators. In this scheme, general formulae may be written for all quantities of interest using W, X, and higher coefficients. Since $D_{10} = D_5 \times C_2$, the relevant W and X follow from Table D4, as discussed for D_6 in Section D5.

The preceding considerations may be applied also to a large number of other aromatic systems. It is interesting to remark that the conclusions of the present section, which were reached purely from the group-theoretic end, are almost identical to those reached by R. G. Parr† in his attempts to find better methods in the semiempirical theory of aromatic systems.

* W. Moffitt, *J. Chem. Phys.*, **22** (1954), 320.
† R. G. Parr, *J. Chem. Phys.*, **33** (1960), 1184.

The interested reader is referred especially to his paper, to the end of Section 5.2 of the present book, and to Section 16c of Fano and Racah (3).

Finally, note that the same techniques could be applied to the free-electron network model of π-electron systems. The presence of cross links here could be dealt with by introducing a perturbation V_{xy} corresponding to a cross link between points x and y on the periphery. The perturbation is a one-electron operator having the matrix elements

$$\langle \bar{\psi} \mid V_{xy} \mid \psi' \rangle = a\beta[\bar{\psi}(x)\psi'(y) + \bar{\psi}(y)\psi'(x)] \qquad (11.11)$$

Here a is the carbon-carbon internuclear distance, and β could be taken as the familiar resonance integral of Hückel theory. The justification of Eq. (11.11) comes from the known relation between the free-electron network model and an algebraic scheme analogous to an *LCAO* model.*

* J. S. Griffith, *Proc. Camb. Phil. Soc.*, **49** (1953), 650, especially Eq. (1).

APPENDIX A

Important Remarks Concerning Phase

The phases of the coupling coefficients for the one-valued representations of the octahedral group are not the same in the present book as my previous one (26). Their signs relative to those in the latter and also to those given earlier by Tanabe and Sugano (20) are in Table A1 below. When either a or b is A_1, the phases are the same in all three places. The phases for the two-valued representations and for the functions of t_2^n and e^n are the same in the two books. The phases for the two-valued representations need some redefinition if generalized V coefficients are defined for them, in analogy with Eq. (3.9) (see ref. 36).

The change of phase is unfortunate, but unavoidable. In Tanabe and Sugano's earlier work, and in mine, there was no compelling reason for making any particular phase choice. However, since the symmetry requirements for the V coefficients fix most of the choices, we must change the phases accordingly.

The present definition of reduced matrix elements for finite group representations [Eq. (2.15)] differs from that in my paper [(25), p. 457] by a factor $(-1)^{a+a'+b}$ but it is essentially the same as that of Tanabe and Kamimura (21). This change is made in order that the definition should exactly parallel that of Fano and Racah for spin functions [Eq. (9.9)]; it is desirable because of Eq. (9.26) relating some of the V and $\overline{V}$.

Table A1. Phases of coupling coefficients. $\langle abc\gamma \mid ab\alpha\beta\rangle$ *given previously by Tanabe and Sugano* (TS) *and myself* (G) *relative to the present book.*

a	*b*	*c*	*TS* (20)	*G* (26)	*a*	*b*	*c*	*TS* (20)	*G* (26)
A_2	A_2	A_1	−	+	T_1	E	T_2	+	−
E	E	A_1	+	+	T_2	E	T_1	+	+
A_2	E	E	−	−	T_2	T_1	E	−	−
E	A_2	E	+	+	E	T_1	T_1	−	−
E	E	A_2	+	+	T_1	E	T_1	−	−
E	E	E	+	+	T_1	T_1	E	+	+
T_1	T_1	A_1	−	+	T_1	T_1	T_1	+	+
T_2	T_2	A_1	+	+	T_1	T_1	T_2	+	+
A_2	T_1	T_2	+	+	T_1	T_2	T_1	−	+
A_2	T_2	T_1	−	+	T_2	T_1	T_1	−	+
T_1	T_2	A_2	−	+	T_1	T_2	T_2	+	+
T_1	A_2	T_2	+	+	T_2	T_1	T_2	−	−
T_2	A_2	T_1	−	+	T_2	T_2	T_1	−	+
T_2	T_1	A_2	−	+	T_2	T_2	T_2	−	+
E	T_1	T_2	−	+	E	T_2	T_2	−	−
E	T_2	T_1	−	−	T_2	E	T_2	−	−
T_1	T_2	E	+	+	T_2	T_2	E	−	+

APPENDIX B

Associated Representations for the Octahedral Group

Here we investigate briefly a further symmetry possessed by V, W, and higher coefficients for the octahedral group. It is of independent interest, but its practical value lies mainly in reducing the number of independent coefficients which have to be calculated to build up tables like C3.1 and C3.2.

We first recall that each irreducible representation a of the octahedral group has associated with it another irreducible representation, a' say, according to the rule $a' = aA_2$. Clearly, $A_1' = A_2$, $A_2' = A_1$, $E' = E$, $T_1' = T_2$, $T_2' = T_1$, and $a'' = a$ always. It also follows from the relations

$$\langle A_2T_1\iota\alpha \mid A_2T_1T_2\beta\rangle = \langle T_1A_2\alpha\iota \mid T_1A_2T_2\beta\rangle = \delta_{\alpha\beta}$$

that if the functions $f_\alpha^{T_1}$ span T_1, and χ transforms as A_2, then $\chi f_\alpha^{T_1}$ is the α component of a set of functions spanning T_2. A similar remark applies on interchanging T_1 and T_2. As a consequence, we may define not only associated representations but also associated components. Thus $(T_1x)' = T_2x$, $(A_1\iota)' = A_2\iota$, etc. The E representation is self-associated, so its behaviour is more complicated. It is discussed later.

We investigate V coefficients by the technique of Section 2.4 and let ϕ, ψ, χ, f_α^a, g_β^b, h_γ^c be real functions transforming respectively as A_1, A_1, A_2, a, b, c. Then χ^2 transforms as A_1, so

$$\langle\phi f_\alpha^a \mid \chi^2 g_\beta^b \mid h_\gamma^c\psi\rangle = \langle\phi f^a \,\|\, \chi^2 g^b \,\|\, h^c\psi\rangle V\begin{pmatrix} a & c & b \\ \alpha & \gamma & \beta\end{pmatrix}$$

$$= \langle\phi\chi f_\alpha^a \mid \chi g_\beta^b \mid h_\gamma^c\psi\rangle = \langle\phi\chi f^a \,\|\, \chi g^b \,\|\, h^c\psi\rangle V\begin{pmatrix} a' & c & b' \\ \alpha' & \gamma & \beta'\end{pmatrix}$$

etc., provided none of a, b, or c is E. Just as in Sections 2.4 and 2.5, we readily deduce that

$$V\begin{pmatrix} a & c & b \\ \alpha & \gamma & \beta \end{pmatrix} = \pm V\begin{pmatrix} a' & c & b' \\ \alpha' & \gamma & \beta' \end{pmatrix} = \pm V\begin{pmatrix} a' & c' & b \\ \alpha' & \gamma' & \beta \end{pmatrix} = \pm V\begin{pmatrix} a & c' & b' \\ \alpha & \gamma' & \beta' \end{pmatrix}$$

with the same set of signs holding for fixed a, b, c independently of α, β, and γ. We use the notation

$$V\begin{pmatrix} a & b & c \\ \alpha & \beta & \gamma \end{pmatrix}'_{\mu\nu} = \omega(abc)_{\mu\nu} V\begin{pmatrix} a & b & c \\ \alpha & \beta & \gamma \end{pmatrix} \tag{B1}$$

where $\omega = \pm 1$ and $\mu\nu$ indicates that the μth and νth columns of V have been "primed." Because of the invariance of V to even permutations, it follows that ω is also unchanged by them, provided that $\mu\nu$ is permuted as well as abc. For example,

$$\omega(abc)_{12} = \omega(bca)_{13} = \omega(cab)_{23}$$

The order of μ and ν is unimportant. Inspection of Tables C2 shows that we have actually arranged things so that

$$\omega(abc)_{\mu\nu} = 1$$

when none of a, b, or c is E (actually this holds for the complex component systems also, although this fact is of no interest to us because W, X, etc. may always be calculated by using a real component system).

The E representation is dealt with by introducing a new pair of components θ' and ϵ' belonging this time to the same representation as do θ and ϵ. We define

$$f^E_{\theta'} = f^E_{\epsilon}, \qquad f^E_{\epsilon'} = -f^E_{\theta} \tag{B2}$$

Then, when we use the coupling coefficients for $A_2\,E$, if χ transforms as A_2, the products χf^E_θ, χf^E_ϵ are the θ', ϵ' components, respectively, of a set of functions transforming as E. Hence, Eq. (B1) holds also when one or more of a, b, or c is E. The only remaining question is how to interpret α', β', and γ' in V coefficients involving E representations. Consider the case $a = E$ and neither b nor c equal to E. Then

$$\langle f^E_{\theta'} \mid g^b_\beta \mid h^c_\gamma\rangle = \langle f^E_{\epsilon} \mid g^b_\beta \mid h^c_\gamma\rangle$$
$$= \langle f^E \,\|\, g^b \,\|\, h^c\rangle V\begin{pmatrix} E & c & b \\ \epsilon & \gamma & \beta \end{pmatrix}$$

We would like the first matrix element to be equal to

$$\langle f^E \,\|\, g^b \,\|\, h^c\rangle V\begin{pmatrix} E & c & b \\ \theta' & \gamma & \beta \end{pmatrix}$$

So we define

$$V\begin{pmatrix} E & c & b \\ \theta' & \gamma & \beta \end{pmatrix} = V\begin{pmatrix} E & c & b \\ \epsilon & \gamma & \beta \end{pmatrix} \tag{B3}$$

and also

$$V\begin{pmatrix} E & c & b \\ \epsilon' & \gamma & \beta \end{pmatrix} = -V\begin{pmatrix} E & c & b \\ \theta & \gamma & \beta \end{pmatrix} \tag{B4}$$

These definitions can be expressed more succinctly by writing $\theta' = \epsilon$ and $\epsilon' = -\theta$, which gives the relationship correctly even when V contains more than one E.

When there are E representations in a V symbol, it is not always possible to choose $\omega = +1$. Take $a = b = c = E$ as an example. Because of the possibility of cyclic permutation,

$$\omega(EEE)_{12} = \omega(EEE)_{13} = \omega(EEE)_{23} = \omega$$

for example. Now

$$V\begin{pmatrix} E & E & E \\ \theta & \theta & \theta \end{pmatrix} = -\tfrac{1}{2}$$

and

$$V\begin{pmatrix} E & E & E \\ \theta & \epsilon & \epsilon \end{pmatrix} = V\begin{pmatrix} E & E & E \\ \epsilon & \theta & \epsilon \end{pmatrix} = V\begin{pmatrix} E & E & E \\ \epsilon & \epsilon & \theta \end{pmatrix} = \tfrac{1}{2}$$

so putting $\alpha = \beta = \gamma = \theta$ in Eq. (B1) gives

$$V\begin{pmatrix} E & E & E \\ \theta & \theta & \theta \end{pmatrix}_{12} = \omega V\begin{pmatrix} E & E & E \\ \theta & \theta & \theta \end{pmatrix} = -\tfrac{1}{2}\omega$$

$$= V\begin{pmatrix} E & E & E \\ \theta' & \theta' & \theta \end{pmatrix} = V\begin{pmatrix} E & E & E \\ \epsilon & \epsilon & \theta \end{pmatrix} = \tfrac{1}{2}$$

and hence $\omega = -1$. ω is equally easily evaluated in other cases and a sufficient table of values is given in Table B1.

Given the ω, we can deduce the relationship between W and W', where the latter has been primed either in two columns or in a set of three symbols which are chosen so that the remaining three symbols occur together in one of the constituent V coefficients in W. For example,

$$W\begin{pmatrix} a & b & c \\ d' & e' & f' \end{pmatrix} = \omega(aef)_{23}\omega(bfd)_{23}\omega(cde)_{23}W\begin{pmatrix} a & b & c \\ d & e & f \end{pmatrix}$$

whence, in particular,

$$W\begin{pmatrix} E & T_1 & T_1 \\ T_2 & T_2 & T_2 \end{pmatrix} = W\begin{pmatrix} E & T_1 & T_1 \\ T_1 & T_1 & T_1 \end{pmatrix}$$

Similar remarks apply to X. Here we may prime any four symbols which are common to a given pair of rows and pair of columns or any six which are chosen so as to omit one symbol from each row and each column. It is evident that $W = W'$ whenever W contains no E. Similarly, $X = X'$ when there is no E in the same row or column as any symbol which has been primed. Thus

$$X\begin{bmatrix} E & T_1 & T_1 \\ T_2 & T_2 & T_2 \\ T_2 & T_2 & T_2 \end{bmatrix} = X\begin{bmatrix} E & T_1 & T_1 \\ T_2 & T_1 & T_1 \\ T_2 & T_1 & T_1 \end{bmatrix}$$

Table B1. Values of $\omega(abc)_{\mu\nu}$.

a	b	c	$\mu\nu$ 12	13	23
E	E	A_2	1	1	−1
E	A_2	E	−1	1	1
A_2	E	E	1	−1	1
E	E	E	−1	−1	−1
E	T_1	T_1	1	−1	1
E	T_1	T_2	1	1	−1
E	T_2	T_1	−1	−1	−1
E	T_2	T_2	−1	1	1

APPENDIX C

Tables for the Octahedral Group

Table C1. Direct products of one-valued representations.

	A_1	A_2	E	T_1	T_2
A_1	A_1	A_2	E	T_1	T_2
A_2	A_2	A_1	E	T_2	T_1
E	E	E	$A_1 + A_2 + E$	$T_1 + T_2$	$T_1 + T_2$
T_1	T_1	T_2	$T_1 + T_2$	$A_1 + E + T_1 + T_2$	$A_2 + E + T_1 + T_2$
T_2	T_2	T_1	$T_1 + T_2$	$A_2 + E + T_1 + T_2$	$A_1 + E + T_1 + T_2$

C2 V COEFFICIENTS. ALL V WHICH ARE NOT GIVEN ARE ZERO.

Table C2.1. Real tetragonal component system.

$$V\begin{pmatrix} A_1 & b & b \\ \iota & \beta & \gamma \end{pmatrix} = \lambda(b)^{-1/2}\delta_{\beta\gamma}, \qquad V\begin{pmatrix} A_2 & T_1 & T_2 \\ \iota & \beta & \gamma \end{pmatrix} = \frac{1}{\sqrt{3}}\,\delta_{\beta\gamma}$$

$$V\begin{pmatrix} T_1 & T_1 & T_1 \\ \alpha & \beta & \gamma \end{pmatrix} = V\begin{pmatrix} T_1 & T_2 & T_2 \\ \alpha & \beta & \gamma \end{pmatrix} = -\frac{1}{\sqrt{6}}\,\epsilon_{\alpha\beta\gamma}$$

$$V\begin{pmatrix} T_1 & T_1 & T_2 \\ \alpha & \beta & \gamma \end{pmatrix} = V\begin{pmatrix} T_2 & T_2 & T_2 \\ \alpha & \beta & \gamma \end{pmatrix} = -\frac{1}{\sqrt{6}}\,|\epsilon_{\alpha\beta\gamma}|$$

A_2	E	E	V
ι	θ	ϵ	$1/\sqrt{2}$
ι	ϵ	θ	$-1/\sqrt{2}$

E	E	E	V
θ	θ	θ	$-\frac{1}{2}$
θ	ϵ	ϵ	$\frac{1}{2}$
ϵ	θ	ϵ	$\frac{1}{2}$
ϵ	ϵ	θ	$\frac{1}{2}$

E	T_1	T_2	V
θ	x	x	$-\frac{1}{2}$
θ	y	y	$\frac{1}{2}$
ϵ	x	x	$-1/2\sqrt{3}$
ϵ	y	y	$-1/2\sqrt{3}$
ϵ	z	z	$1/\sqrt{3}$

E T_1 T_1 or E T_2 T_2			V
θ	x	x	$1/2\sqrt{3}$
θ	y	y	$1/2\sqrt{3}$
θ	z	z	$-1/\sqrt{3}$
ϵ	x	x	$-\frac{1}{2}$
ϵ	y	y	$\frac{1}{2}$

Table C2.2. Real trigonal component system. V for A_2E^2, E^3, $A_2T_1T_2$, T_1^3, $T_1T_2^2$ and all those including A_1 are as in C2.1.

E	T_1	T_2	V
θ	τ	σ	$-1/\sqrt{6}$
θ	σ	τ	$-1/\sqrt{6}$
θ	ρ	σ	$1/2\sqrt{3}$
θ	σ	ρ	$1/2\sqrt{3}$
ϵ	ρ	ρ	$1/2\sqrt{3}$
ϵ	σ	σ	$-1/2\sqrt{3}$
ϵ	ρ	τ	$1/\sqrt{6}$
ϵ	τ	ρ	$1/\sqrt{6}$

T_1 or T_2	T_1 or T_2	T_2 or T_2	V
τ	τ	τ	$-\sqrt{2}/3$
ρ	ρ	ρ	$-\frac{1}{3}$
τ	ρ	ρ	$\frac{1}{3\sqrt{2}}$
ρ	τ	ρ	
ρ	ρ	τ	
τ	σ	σ	$\frac{1}{3\sqrt{2}}$
σ	τ	σ	
σ	σ	τ	
ρ	σ	σ	$\frac{1}{3}$
σ	ρ	σ	
σ	σ	ρ	

E or E	T_1 or T_2	T_1 or T_2	V
θ	ρ	ρ	$-1/2\sqrt{3}$
θ	σ	σ	$1/2\sqrt{3}$
θ	ρ	τ	$-1/\sqrt{6}$
θ	τ	ρ	$-1/\sqrt{6}$
ϵ	τ	σ	$-1/\sqrt{6}$
ϵ	σ	τ	$-1/\sqrt{6}$
ϵ	ρ	σ	$1/2\sqrt{3}$
ϵ	σ	ρ	$1/2\sqrt{3}$

Table C2.3. Complex tetragonal component system. Components ordered 1, 0, −1. All those not involving T_1 or T_2 are as under Table C2.1.

$$V\begin{pmatrix} T_1 & T_1 & T_1 \\ \alpha & \beta & \gamma \end{pmatrix} = V\begin{pmatrix} T_1 & T_2 & T_2 \\ \alpha & \beta & \gamma \end{pmatrix} = \frac{1}{\sqrt{6}}\epsilon_{\alpha\beta\gamma}$$

A_1 T_1 T_1 or A_1 T_2 T_2 or A_2 T_1 T_2			V
ι	1	−1	$1/\sqrt{3}$
ι	0	0	$-1/\sqrt{3}$
ι	−1	1	$1/\sqrt{3}$

E T_1 T_1 or E T_2 T_2			V
θ	1	−1	$1/2\sqrt{3}$
θ	0	0	$1/\sqrt{3}$
θ	−1	1	$1/2\sqrt{3}$
ϵ	1	1	$\frac{1}{2}$
ϵ	−1	−1	$\frac{1}{2}$

T_1 T_1 T_2 or T_2 T_2 T_2			V
1	1	0	$-\frac{1}{\sqrt{6}}$
1	0	1	
0	1	1	
0	−1	−1	$\frac{1}{\sqrt{6}}$
−1	0	−1	
−1	−1	0	

E	T_1	T_2	V
θ	1	1	$\frac{1}{2}$
θ	−1	−1	$\frac{1}{2}$
ϵ	1	−1	$-1/2\sqrt{3}$
ϵ	0	0	$-1/\sqrt{3}$
ϵ	−1	1	$-1/2\sqrt{3}$

Table C2.4. Complex trigonal component system. Components ordered as 1, 0, −1. A_1^3 and $A_1A_2^2$ as under C2.1.

$$V \begin{pmatrix} T_1 & T_1 & T_1 \\ \alpha & \beta & \gamma \end{pmatrix} = V \begin{pmatrix} T_1 & T_2 & T_2 \\ \alpha & \beta & \gamma \end{pmatrix} = \frac{1}{\sqrt{6}} |\epsilon_{\alpha\beta\gamma}|$$

	A_1	T_1	T_1	V
or	A_1	T_2	T_2	
or	A_2	T_1	T_2	
	ι	1	−1	$1/\sqrt3$
	ι	0	0	$-1/\sqrt3$
	ι	−1	1	$1/\sqrt3$

				V		
					E	
E	E	A_1	A_2	1	−1
1	1	0	0	$-i/\sqrt2$	0
1	−1	$1/\sqrt2$	$-i/\sqrt2$	0	0
−1	1	$1/\sqrt2$	$i/\sqrt2$	0	0
−1	−1	0	0	0	$i/\sqrt2$

	E	T_1	T_1	V
or	E	T_2	T_2	
	1	1	1	$-i/\sqrt6$
	−1	−1	−1	$i/\sqrt6$
	−1	1	0	
	−1	0	1	
	1	−1	0	$-i/\sqrt6$
	1	0	−1	

	T_1	T_1	T_2	V
or	T_2	T_2	T_2	
	0	0	0	$i\sqrt2/3$
	1	1	1	$-i\sqrt2/3$
	−1	−1	−1	$i\sqrt2/3$
	0	1	−1	
	0	−1	1	
	1	0	−1	$\dfrac{i}{3\sqrt2}$
	−1	0	1	
	1	−1	0	
	−1	1	0	

E	T_1	T_2	V
−1	1	−1	$i/\sqrt6$
1	−1	1	$-i/\sqrt6$
−1	0	1	$1/\sqrt6$
1	0	−1	$-1/\sqrt6$
−1	1	0	$1/\sqrt6$
1	−1	0	$-1/\sqrt6$

C3 W AND X COEFFICIENTS

Both tables are partially ordered in the following manner. A three-digit number x is formed whose first, second, and third digits are, respectively, the number of times the representation A_2, E, and T_2 occurs in the coefficient. The coefficients are in an order of increasing x, first along rows and then down columns. Thus, for example, $x = 124$ for

$$X\begin{bmatrix} A_2 & T_1 & T_2 \\ T_1 & E & T_2 \\ T_2 & T_2 & E \end{bmatrix}$$

In both tables, T_1 is represented as a dot and the symbol W or X is omitted. In the table of X, the value of x for the X in the left-hand column is given, to aid in location.

All W containing A_1 are to be evaluated by using Eq. (4.2). Every other W [Eq. (4.1)] either is trivially zero because

$$\delta(a, b, c)\delta(a, e, f)\delta(b, f, d)\delta(c, d, e) = 0$$

or can be rearranged by one of the symmetry operations mentioned in the first paragraph of Section 4.2 to give one of the W in Table C3.1.

All X containing A_1 are to be reduced to a W by using Eq. (8.8), and the W are to be obtained as just described. The X [of Eq. (8.1)] which are trivially zero because Eq. (8.2) is not satisfied are omitted. Any other X may be turned into one given in Table C3.2, or into one containing A_1, by a combination of two procedures. The first is one of the symmetry operations described in Section 8.1, the second paragraph. The other is the application of one of the symmetry operations mentioned at the end of Appendix B; that is, the "priming" of X in such a way that no E representation is in the same row or column of X as is any symbol which becomes "primed." Such an operation leaves the value of X unaltered.

Table C3.1. Table of W. All W with six constituent T representations are equal to $\frac{1}{6}$.

$\begin{pmatrix} E & \cdot & \cdot \\ \cdot & \cdot & \cdot \end{pmatrix} = \frac{1}{6}$	$\begin{pmatrix} E & \cdot & \cdot \\ \cdot & \cdot & T_2 \end{pmatrix} = -\frac{1}{2\sqrt{3}}$	$\begin{pmatrix} E & \cdot & \cdot \\ T_2 & \cdot & \cdot \end{pmatrix} = -\frac{1}{6}$
$\begin{pmatrix} E & \cdot & \cdot \\ \cdot & T_2 & T_2 \end{pmatrix} = -\frac{1}{6}$	$\begin{pmatrix} E & \cdot & \cdot \\ T_2 & \cdot & T_2 \end{pmatrix} = \frac{1}{2\sqrt{3}}$	$\begin{pmatrix} E & \cdot & T_2 \\ \cdot & \cdot & T_2 \end{pmatrix} = -\frac{1}{6}$
$\begin{pmatrix} E & \cdot & T_2 \\ \cdot & T_2 & \cdot \end{pmatrix} = -\frac{1}{6}$	$\begin{pmatrix} E & \cdot & \cdot \\ T_2 & T_2 & T_2 \end{pmatrix} = \frac{1}{6}$	$\begin{pmatrix} E & \cdot & T_2 \\ \cdot & T_2 & T_2 \end{pmatrix} = \frac{1}{2\sqrt{3}}$
$\begin{pmatrix} E & \cdot & T_2 \\ T_2 & \cdot & T_2 \end{pmatrix} = \frac{1}{6}$	$\begin{pmatrix} E & \cdot & T_2 \\ T_2 & T_2 & \cdot \end{pmatrix} = \frac{1}{6}$	$\begin{pmatrix} E & \cdot & T_2 \\ T_2 & T_2 & T_2 \end{pmatrix} = -\frac{1}{2\sqrt{3}}$
$\begin{pmatrix} E & T_2 & T_2 \\ \cdot & T_2 & T_2 \end{pmatrix} = \frac{1}{6}$	$\begin{pmatrix} E & T_2 & T_2 \\ T_2 & T_2 & T_2 \end{pmatrix} = -\frac{1}{6}$	$\begin{pmatrix} E & \cdot & \cdot \\ E & \cdot & \cdot \end{pmatrix} = \frac{1}{3}$
$\begin{pmatrix} E & \cdot & \cdot \\ E & \cdot & T_2 \end{pmatrix} = 0$	$\begin{pmatrix} E & \cdot & \cdot \\ E & T_2 & T_2 \end{pmatrix} = -\frac{1}{3}$	$\begin{pmatrix} E & \cdot & T_2 \\ E & \cdot & T_2 \end{pmatrix} = \frac{1}{3}$
$\begin{pmatrix} E & \cdot & T_2 \\ E & T_2 & \cdot \end{pmatrix} = -\frac{1}{3}$	$\begin{pmatrix} E & \cdot & T_2 \\ E & T_2 & T_2 \end{pmatrix} = 0$	$\begin{pmatrix} E & T_2 & T_2 \\ E & T_2 & T_2 \end{pmatrix} = \frac{1}{3}$
$\begin{pmatrix} E & E & E \\ \cdot & \cdot & \cdot \end{pmatrix} = \frac{1}{2\sqrt{3}}$	$\begin{pmatrix} E & E & E \\ \cdot & \cdot & T_2 \end{pmatrix} = \frac{1}{2\sqrt{3}}$	$\begin{pmatrix} E & E & E \\ \cdot & T_2 & T_2 \end{pmatrix} = \frac{1}{2\sqrt{3}}$
$\begin{pmatrix} E & E & E \\ T_2 & T_2 & T_2 \end{pmatrix} = \frac{1}{2\sqrt{3}}$	$\begin{pmatrix} E & E & E \\ E & E & E \end{pmatrix} = 0$	$\begin{pmatrix} A_2 & \cdot & T_2 \\ \cdot & \cdot & T_2 \end{pmatrix} = \frac{1}{3}$
$\begin{pmatrix} A_2 & \cdot & T_2 \\ \cdot & T_2 & \cdot \end{pmatrix} = -\frac{1}{3}$	$\begin{pmatrix} A_2 & \cdot & T_2 \\ T_2 & \cdot & T_2 \end{pmatrix} = -\frac{1}{3}$	$\begin{pmatrix} A_2 & \cdot & T_2 \\ T_2 & T_2 & \cdot \end{pmatrix} = \frac{1}{3}$
$\begin{pmatrix} A_2 & \cdot & T_2 \\ E & \cdot & T_2 \end{pmatrix} = \frac{1}{3}$	$\begin{pmatrix} A_2 & \cdot & T_2 \\ E & T_2 & \cdot \end{pmatrix} = \frac{1}{3}$	$\begin{pmatrix} A_2 & E & E \\ \cdot & \cdot & T_2 \end{pmatrix} = -\frac{1}{\sqrt{6}}$
$\begin{pmatrix} A_2 & E & E \\ T_2 & \cdot & T_2 \end{pmatrix} = \frac{1}{\sqrt{6}}$	$\begin{pmatrix} A_2 & E & E \\ E & E & E \end{pmatrix} = \frac{1}{2}$	$\begin{pmatrix} A_2 & \cdot & T_2 \\ A_2 & \cdot & T_2 \end{pmatrix} = \frac{1}{3}$
$\begin{pmatrix} A_2 & E & E \\ A_2 & E & E \end{pmatrix} = \frac{1}{2}$		

Table C3.2. Table of X.

000	$\begin{bmatrix} \cdot & \cdot & \cdot \\ \cdot & \cdot & \cdot \\ \cdot & \cdot & \cdot \end{bmatrix} = 0$	$\begin{bmatrix} \cdot & \cdot & \cdot \\ \cdot & \cdot & \cdot \\ \cdot & \cdot & T_2 \end{bmatrix} = \frac{1}{18}$	$\begin{bmatrix} \cdot & \cdot & \cdot \\ \cdot & \cdot & \cdot \\ \cdot & T_2 & T_2 \end{bmatrix} = 0$
002	$\begin{bmatrix} \cdot & \cdot & \cdot \\ \cdot & \cdot & T_2 \\ \cdot & T_2 & \cdot \end{bmatrix} = 0$	$\begin{bmatrix} \cdot & \cdot & \cdot \\ \cdot & \cdot & \cdot \\ T_2 & T_2 & T_2 \end{bmatrix} = \frac{1}{18}$	$\begin{bmatrix} \cdot & \cdot & T_2 \\ \cdot & T_2 & \cdot \\ T_2 & \cdot & \cdot \end{bmatrix} = \frac{1}{18}$
010	$\begin{bmatrix} E & \cdot & \cdot \\ \cdot & \cdot & \cdot \\ \cdot & \cdot & \cdot \end{bmatrix} = \frac{1}{18}$	$\begin{bmatrix} E & \cdot & \cdot \\ \cdot & \cdot & \cdot \\ \cdot & \cdot & T_2 \end{bmatrix} = \frac{1}{18}$	$\begin{bmatrix} E & \cdot & \cdot \\ \cdot & \cdot & \cdot \\ T_2 & \cdot & \cdot \end{bmatrix} = 0$
012	$\begin{bmatrix} E & \cdot & \cdot \\ \cdot & \cdot & \cdot \\ \cdot & T_2 & T_2 \end{bmatrix} = -\frac{1}{18}$	$\begin{bmatrix} E & \cdot & \cdot \\ \cdot & \cdot & \cdot \\ T_2 & \cdot & T_2 \end{bmatrix} = 0$	$\begin{bmatrix} E & \cdot & \cdot \\ \cdot & \cdot & T_2 \\ \cdot & T_2 & \cdot \end{bmatrix} = \frac{1}{18}$
012	$\begin{bmatrix} E & \cdot & \cdot \\ \cdot & \cdot & T_2 \\ T_2 & \cdot & \cdot \end{bmatrix} = 0$	$\begin{bmatrix} E & \cdot & \cdot \\ T_2 & \cdot & \cdot \\ T_2 & \cdot & \cdot \end{bmatrix} = -\frac{1}{18}$	$\begin{bmatrix} E & \cdot & T_2 \\ \cdot & \cdot & \cdot \\ T_2 & \cdot & \cdot \end{bmatrix} = \frac{1}{18}$
013	$\begin{bmatrix} E & \cdot & \cdot \\ \cdot & \cdot & \cdot \\ T_2 & T_2 & T_2 \end{bmatrix} = 0$	$\begin{bmatrix} E & \cdot & \cdot \\ \cdot & \cdot & T_2 \\ T_2 & T_2 & \cdot \end{bmatrix} = 0$	$\begin{bmatrix} E & \cdot & \cdot \\ T_2 & \cdot & \cdot \\ T_2 & \cdot & T_2 \end{bmatrix} = -\frac{1}{18}$
013	$\begin{bmatrix} E & \cdot & \cdot \\ T_2 & \cdot & T_2 \\ \cdot & \cdot & T_2 \end{bmatrix} = 0$	$\begin{bmatrix} E & \cdot & T_2 \\ \cdot & \cdot & \cdot \\ T_2 & \cdot & T_2 \end{bmatrix} = \frac{1}{18}$	$\begin{bmatrix} E & \cdot & T_2 \\ \cdot & \cdot & \cdot \\ T_2 & T_2 & \cdot \end{bmatrix} = -\frac{1}{18}$
013	$\begin{bmatrix} E & \cdot & T_2 \\ \cdot & T_2 & \cdot \\ T_2 & \cdot & \cdot \end{bmatrix} = \frac{1}{18}$	$\begin{bmatrix} E & \cdot & T_2 \\ T_2 & \cdot & \cdot \\ T_2 & \cdot & \cdot \end{bmatrix} = 0$	$\begin{bmatrix} E & \cdot & \cdot \\ T_2 & \cdot & \cdot \\ T_2 & T_2 & T_2 \end{bmatrix} = \frac{1}{18}$
014	$\begin{bmatrix} E & \cdot & \cdot \\ T_2 & \cdot & T_2 \\ T_2 & \cdot & T_2 \end{bmatrix} = \frac{1}{18}$	$\begin{bmatrix} E & \cdot & \cdot \\ T_2 & \cdot & T_2 \\ T_2 & T_2 & \cdot \end{bmatrix} = -\frac{1}{18}$	$\begin{bmatrix} E & \cdot & T_2 \\ \cdot & \cdot & \cdot \\ T_2 & T_2 & T_2 \end{bmatrix} = -\frac{1}{18}$
014	$\begin{bmatrix} E & \cdot & T_2 \\ \cdot & \cdot & T_2 \\ T_2 & T_2 & \cdot \end{bmatrix} = \frac{1}{18}$	$\begin{bmatrix} E & \cdot & T_2 \\ T_2 & \cdot & \cdot \\ T_2 & \cdot & T_2 \end{bmatrix} = 0$	$\begin{bmatrix} E & \cdot & T_2 \\ T_2 & \cdot & \cdot \\ T_2 & T_2 & \cdot \end{bmatrix} = 0$
014	$\begin{bmatrix} E & T_2 & T_2 \\ T_2 & \cdot & \cdot \\ T_2 & \cdot & \cdot \end{bmatrix} = \frac{1}{18}$	$\begin{bmatrix} E & \cdot & T_2 \\ T_2 & \cdot & \cdot \\ T_2 & T_2 & T_2 \end{bmatrix} = 0$	$\begin{bmatrix} E & \cdot & T_2 \\ T_2 & \cdot & T_2 \\ T_2 & \cdot & T_2 \end{bmatrix} = 0$
015	$\begin{bmatrix} E & \cdot & T_2 \\ T_2 & \cdot & T_2 \\ T_2 & T_2 & \cdot \end{bmatrix} = 0$	$\begin{bmatrix} E & T_2 & T_2 \\ T_2 & \cdot & \cdot \\ T_2 & \cdot & T_2 \end{bmatrix} = \frac{1}{18}$	$\begin{bmatrix} E & T_2 & T_2 \\ T_2 & \cdot & \cdot \\ T_2 & T_2 & T_2 \end{bmatrix} = -\frac{1}{18}$
016	$\begin{bmatrix} E & T_2 & T_2 \\ T_2 & \cdot & T_2 \\ T_2 & T_2 & \cdot \end{bmatrix} = \frac{1}{18}$	$\begin{bmatrix} E & \cdot & \cdot \\ \cdot & E & \cdot \\ \cdot & \cdot & \cdot \end{bmatrix} = -\frac{1}{36}$	$\begin{bmatrix} E & \cdot & \cdot \\ \cdot & E & \cdot \\ \cdot & \cdot & T_2 \end{bmatrix} = \frac{1}{36}$

Table C3.2. Table of X. (cont'd)

021	$\begin{bmatrix} E & \cdot & \cdot \\ \cdot & E & \cdot \\ \cdot & T_2 & \cdot \end{bmatrix} = -\frac{1}{12\sqrt{3}}$	$\begin{bmatrix} E & \cdot & \cdot \\ T_2 & E & \cdot \\ \cdot & \cdot & \cdot \end{bmatrix} = \frac{1}{12}$	$\begin{bmatrix} E & \cdot & \cdot \\ \cdot & E & \cdot \\ \cdot & T_2 & T_2 \end{bmatrix} = \frac{1}{12\sqrt{3}}$
022	$\begin{bmatrix} E & \cdot & \cdot \\ \cdot & E & \cdot \\ T_2 & T_2 & \cdot \end{bmatrix} = -\frac{1}{12}$	$\begin{bmatrix} E & \cdot & \cdot \\ \cdot & E & T_2 \\ \cdot & T_2 & \cdot \end{bmatrix} = \frac{1}{36}$	$\begin{bmatrix} E & \cdot & \cdot \\ \cdot & E & T_2 \\ T_2 & \cdot & \cdot \end{bmatrix} = -\frac{1}{12}$
022	$\begin{bmatrix} E & \cdot & \cdot \\ T_2 & E & \cdot \\ \cdot & \cdot & T_2 \end{bmatrix} = -\frac{1}{12}$	$\begin{bmatrix} E & \cdot & \cdot \\ T_2 & E & \cdot \\ \cdot & T_2 & \cdot \end{bmatrix} = -\frac{1}{12\sqrt{3}}$	$\begin{bmatrix} E & \cdot & \cdot \\ T_2 & E & \cdot \\ T_2 & \cdot & \cdot \end{bmatrix} = -\frac{1}{12\sqrt{3}}$
022	$\begin{bmatrix} E & \cdot & \cdot \\ T_2 & E & T_2 \\ \cdot & \cdot & \cdot \end{bmatrix} = -\frac{1}{12\sqrt{3}}$	$\begin{bmatrix} E & \cdot & T_2 \\ T_2 & E & \cdot \\ \cdot & \cdot & \cdot \end{bmatrix} = -\frac{1}{12\sqrt{3}}$	$\begin{bmatrix} E & T_2 & \cdot \\ T_2 & E & \cdot \\ \cdot & \cdot & \cdot \end{bmatrix} = -\frac{1}{36}$
023	$\begin{bmatrix} E & \cdot & \cdot \\ \cdot & E & \cdot \\ T_2 & T_2 & T_2 \end{bmatrix} = \frac{1}{12}$	$\begin{bmatrix} E & \cdot & \cdot \\ \cdot & E & T_2 \\ \cdot & T_2 & T_2 \end{bmatrix} = -\frac{1}{36}$	$\begin{bmatrix} E & \cdot & \cdot \\ \cdot & E & T_2 \\ T_2 & \cdot & T_2 \end{bmatrix} = \frac{1}{12}$
023	$\begin{bmatrix} E & \cdot & \cdot \\ \cdot & E & T_2 \\ T_2 & T_2 & \cdot \end{bmatrix} = \frac{1}{12\sqrt{3}}$	$\begin{bmatrix} E & \cdot & \cdot \\ T_2 & E & \cdot \\ \cdot & T_2 & T_2 \end{bmatrix} = \frac{1}{12\sqrt{3}}$	$\begin{bmatrix} E & \cdot & \cdot \\ T_2 & E & \cdot \\ T_2 & T_2 & \cdot \end{bmatrix} = \frac{1}{36}$
023	$\begin{bmatrix} E & \cdot & \cdot \\ T_2 & E & T_2 \\ \cdot & \cdot & T_2 \end{bmatrix} = \frac{1}{12\sqrt{3}}$	$\begin{bmatrix} E & \cdot & \cdot \\ T_2 & E & T_2 \\ \cdot & T_2 & \cdot \end{bmatrix} = -\frac{1}{12}$	$\begin{bmatrix} E & \cdot & \cdot \\ T_2 & E & T_2 \\ T_2 & \cdot & \cdot \end{bmatrix} = \frac{1}{36}$
023	$\begin{bmatrix} E & \cdot & T_2 \\ T_2 & E & \cdot \\ \cdot & T_2 & \cdot \end{bmatrix} = \frac{1}{36}$	$\begin{bmatrix} E & \cdot & T_2 \\ T_2 & E & T_2 \\ \cdot & \cdot & \cdot \end{bmatrix} = \frac{1}{36}$	$\begin{bmatrix} E & T_2 & \cdot \\ T_2 & E & \cdot \\ \cdot & \cdot & T_2 \end{bmatrix} = \frac{1}{36}$
023	$\begin{bmatrix} E & T_2 & \cdot \\ T_2 & E & \cdot \\ \cdot & T_2 & \cdot \end{bmatrix} = -\frac{1}{12\sqrt{3}}$	$\begin{bmatrix} E & \cdot & \cdot \\ \cdot & E & T_2 \\ T_2 & T_2 & T_2 \end{bmatrix} = -\frac{1}{12\sqrt{3}}$	$\begin{bmatrix} E & \cdot & \cdot \\ T_2 & E & \cdot \\ T_2 & T_2 & T_2 \end{bmatrix} = -\frac{1}{36}$
024	$\begin{bmatrix} E & \cdot & \cdot \\ T_2 & E & T_2 \\ \cdot & T_2 & T_2 \end{bmatrix} = \frac{1}{12}$	$\begin{bmatrix} E & \cdot & \cdot \\ T_2 & E & T_2 \\ T_2 & \cdot & T_2 \end{bmatrix} = -\frac{1}{36}$	$\begin{bmatrix} E & \cdot & \cdot \\ T_2 & E & T_2 \\ T_2 & T_2 & \cdot \end{bmatrix} = \frac{1}{12\sqrt{3}}$
024	$\begin{bmatrix} E & \cdot & T_2 \\ \cdot & E & T_2 \\ T_2 & T_2 & \cdot \end{bmatrix} = -\frac{1}{36}$	$\begin{bmatrix} E & \cdot & T_2 \\ T_2 & E & \cdot \\ \cdot & T_2 & T_2 \end{bmatrix} = -\frac{1}{36}$	$\begin{bmatrix} E & \cdot & T_2 \\ T_2 & E & T_2 \\ \cdot & \cdot & T_2 \end{bmatrix} = -\frac{1}{36}$
024	$\begin{bmatrix} E & \cdot & T_2 \\ T_2 & E & T_2 \\ \cdot & T_2 & \cdot \end{bmatrix} = \frac{1}{12\sqrt{3}}$	$\begin{bmatrix} E & T_2 & \cdot \\ T_2 & E & \cdot \\ \cdot & T_2 & T_2 \end{bmatrix} = \frac{1}{12\sqrt{3}}$	$\begin{bmatrix} E & T_2 & \cdot \\ T_2 & E & \cdot \\ T_2 & T_2 & \cdot \end{bmatrix} = -\frac{1}{12}$
024	$\begin{bmatrix} E & T_2 & \cdot \\ T_2 & E & T_2 \\ \cdot & T_2 & \cdot \end{bmatrix} = \frac{1}{36}$	$\begin{bmatrix} E & T_2 & \cdot \\ T_2 & E & T_2 \\ T_2 & \cdot & \cdot \end{bmatrix} = -\frac{1}{12}$	$\begin{bmatrix} E & \cdot & \cdot \\ T_2 & E & T_2 \\ T_2 & T_2 & T_2 \end{bmatrix} = -\frac{1}{12\sqrt{3}}$

Table C3.2. Table of X. (cont'd)

025	$\begin{bmatrix} E & \cdot & T_2 \\ \cdot & E & T_2 \\ T_2 & T_2 & T_2 \end{bmatrix} = \frac{1}{36}$	$\begin{bmatrix} E & \cdot & T_2 \\ T_2 & E & T_2 \\ \cdot & T_2 & T_2 \end{bmatrix} = -\frac{1}{12\sqrt{3}}$	$\begin{bmatrix} E & \cdot & T_2 \\ T_2 & E & T_2 \\ T_2 & T_2 & \cdot \end{bmatrix} = \frac{1}{12}$
025	$\begin{bmatrix} E & T_2 & \cdot \\ T_2 & E & \cdot \\ T_2 & T_2 & T_2 \end{bmatrix} = \frac{1}{12}$	$\begin{bmatrix} E & T_2 & \cdot \\ T_2 & E & T_2 \\ \cdot & T_2 & T_2 \end{bmatrix} = -\frac{1}{36}$	$\begin{bmatrix} E & T_2 & \cdot \\ T_2 & E & T_2 \\ T_2 & \cdot & T_2 \end{bmatrix} = \frac{1}{12}$
025	$\begin{bmatrix} E & T_2 & \cdot \\ T_2 & E & T_2 \\ T_2 & T_2 & \cdot \end{bmatrix} = \frac{1}{12\sqrt{3}}$	$\begin{bmatrix} E & T_2 & T_2 \\ T_2 & E & \cdot \\ T_2 & T_2 & \cdot \end{bmatrix} = \frac{1}{12\sqrt{3}}$	$\begin{bmatrix} E & \cdot & T_2 \\ T_2 & E & T_2 \\ T_2 & T_2 & T_2 \end{bmatrix} = -\frac{1}{12}$
026	$\begin{bmatrix} E & T_2 & \cdot \\ T_2 & E & T_2 \\ T_2 & T_2 & T_2 \end{bmatrix} = -\frac{1}{12\sqrt{3}}$	$\begin{bmatrix} E & T_2 & T_2 \\ T_2 & E & T_2 \\ T_2 & T_2 & \cdot \end{bmatrix} = -\frac{1}{36}$	$\begin{bmatrix} E & T_2 & T_2 \\ T_2 & E & T_2 \\ T_2 & T_2 & T_2 \end{bmatrix} = \frac{1}{36}$
030	$\begin{bmatrix} E & E & E \\ \cdot & \cdot & \cdot \\ \cdot & \cdot & \cdot \end{bmatrix} = \frac{1}{6\sqrt{3}}$	$\begin{bmatrix} E & \cdot & \cdot \\ \cdot & E & \cdot \\ \cdot & \cdot & E \end{bmatrix} = \frac{1}{9}$	$\begin{bmatrix} E & E & E \\ \cdot & \cdot & \cdot \\ \cdot & \cdot & T_2 \end{bmatrix} = 0$
031	$\begin{bmatrix} E & \cdot & \cdot \\ \cdot & E & \cdot \\ \cdot & T_2 & E \end{bmatrix} = 0$	$\begin{bmatrix} E & E & E \\ \cdot & \cdot & \cdot \\ \cdot & T_2 & T_2 \end{bmatrix} = -\frac{1}{6\sqrt{3}}$	$\begin{bmatrix} E & E & E \\ \cdot & \cdot & T_2 \\ \cdot & \cdot & T_2 \end{bmatrix} = \frac{1}{6\sqrt{3}}$
032	$\begin{bmatrix} E & E & E \\ \cdot & \cdot & T_2 \\ \cdot & T_2 & \cdot \end{bmatrix} = \frac{1}{6\sqrt{3}}$	$\begin{bmatrix} E & \cdot & \cdot \\ \cdot & E & \cdot \\ T_2 & T_2 & E \end{bmatrix} = 0$	$\begin{bmatrix} E & \cdot & \cdot \\ \cdot & E & T_2 \\ \cdot & T_2 & E \end{bmatrix} = \frac{1}{9}$
032	$\begin{bmatrix} E & \cdot & \cdot \\ T_2 & E & \cdot \\ \cdot & T_2 & E \end{bmatrix} = 0$	$\begin{bmatrix} E & E & E \\ \cdot & \cdot & \cdot \\ T_2 & T_2 & T_2 \end{bmatrix} = 0$	$\begin{bmatrix} E & E & E \\ \cdot & \cdot & T_2 \\ \cdot & T_2 & T_2 \end{bmatrix} = 0$
033	$\begin{bmatrix} E & E & E \\ \cdot & \cdot & T_2 \\ T_2 & T_2 & \cdot \end{bmatrix} = 0$	$\begin{bmatrix} E & \cdot & \cdot \\ \cdot & E & T_2 \\ T_2 & T_2 & E \end{bmatrix} = 0$	$\begin{bmatrix} E & \cdot & \cdot \\ T_2 & E & \cdot \\ T_2 & T_2 & E \end{bmatrix} = -\frac{1}{9}$
033	$\begin{bmatrix} E & \cdot & \cdot \\ T_2 & E & T_2 \\ T_2 & \cdot & E \end{bmatrix} = -\frac{1}{9}$	$\begin{bmatrix} E & \cdot & T_2 \\ T_2 & E & \cdot \\ \cdot & T_2 & E \end{bmatrix} = -\frac{1}{9}$	$\begin{bmatrix} E & E & E \\ \cdot & \cdot & T_2 \\ T_2 & T_2 & T_2 \end{bmatrix} = -\frac{1}{6\sqrt{3}}$
034	$\begin{bmatrix} E & E & E \\ \cdot & T_2 & T_2 \\ \cdot & T_2 & T_2 \end{bmatrix} = \frac{1}{6\sqrt{3}}$	$\begin{bmatrix} E & E & E \\ \cdot & T_2 & T_2 \\ T_2 & \cdot & T_2 \end{bmatrix} = \frac{1}{6\sqrt{3}}$	$\begin{bmatrix} E & \cdot & \cdot \\ T_2 & E & T_2 \\ T_2 & T_2 & E \end{bmatrix} = 0$
034	$\begin{bmatrix} E & \cdot & T_2 \\ \cdot & E & T_2 \\ T_2 & T_2 & E \end{bmatrix} = \frac{1}{9}$	$\begin{bmatrix} E & \cdot & T_2 \\ T_2 & E & \cdot \\ T_2 & T_2 & E \end{bmatrix} = 0$	$\begin{bmatrix} E & E & E \\ \cdot & T_2 & T_2 \\ T_2 & T_2 & T_2 \end{bmatrix} = 0$
035	$\begin{bmatrix} E & \cdot & T_2 \\ T_2 & E & T_2 \\ T_2 & T_2 & E \end{bmatrix} = 0$	$\begin{bmatrix} E & E & E \\ T_2 & T_2 & T_2 \\ T_2 & T_2 & T_2 \end{bmatrix} = \frac{1}{6\sqrt{3}}$	$\begin{bmatrix} E & T_2 & T_2 \\ T_2 & E & T_2 \\ T_2 & T_2 & E \end{bmatrix} = \frac{1}{9}$

Table C3.2. Table of X. (cont'd)

050	$\begin{bmatrix} E & E & E \\ E & \cdot & \cdot \\ E & \cdot & \cdot \end{bmatrix} = \frac{1}{12}$	$\begin{bmatrix} E & E & E \\ E & \cdot & \cdot \\ E & \cdot & T_2 \end{bmatrix} = \frac{1}{12}$	$\begin{bmatrix} E & E & E \\ E & \cdot & \cdot \\ E & T_2 & T_2 \end{bmatrix} = -\frac{1}{12}$
052	$\begin{bmatrix} E & E & E \\ E & \cdot & T_2 \\ E & T_2 & \cdot \end{bmatrix} = \frac{1}{12}$	$\begin{bmatrix} E & E & E \\ E & \cdot & T_2 \\ E & T_2 & T_2 \end{bmatrix} = \frac{1}{12}$	$\begin{bmatrix} E & E & E \\ E & T_2 & T_2 \\ E & T_2 & T_2 \end{bmatrix} = \frac{1}{12}$
090	$\begin{bmatrix} E & E & E \\ E & E & E \\ E & E & E \end{bmatrix} = \frac{1}{4}$	$\begin{bmatrix} A_2 & E & E \\ \cdot & \cdot & \cdot \\ T_2 & \cdot & \cdot \end{bmatrix} = \frac{1}{6\sqrt{2}}$	$\begin{bmatrix} A_2 & E & E \\ \cdot & \cdot & \cdot \\ T_2 & \cdot & T_2 \end{bmatrix} = \frac{1}{6\sqrt{6}}$
122	$\begin{bmatrix} A_2 & E & E \\ \cdot & \cdot & T_2 \\ T_2 & \cdot & \cdot \end{bmatrix} = -\frac{1}{6\sqrt{6}}$	$\begin{bmatrix} A_2 & \cdot & T_2 \\ \cdot & E & \cdot \\ T_2 & \cdot & E \end{bmatrix} = \frac{1}{9}$	$\begin{bmatrix} A_2 & \cdot & T_2 \\ T_2 & E & \cdot \\ \cdot & \cdot & E \end{bmatrix} = 0$
123	$\begin{bmatrix} A_2 & E & E \\ \cdot & \cdot & \cdot \\ T_2 & T_2 & T_2 \end{bmatrix} = \frac{1}{6\sqrt{2}}$	$\begin{bmatrix} A_2 & E & E \\ \cdot & \cdot & T_2 \\ T_2 & \cdot & T_2 \end{bmatrix} = \frac{1}{6\sqrt{2}}$	$\begin{bmatrix} A_2 & E & E \\ \cdot & \cdot & T_2 \\ T_2 & T_2 & \cdot \end{bmatrix} = -\frac{1}{6\sqrt{2}}$
123	$\begin{bmatrix} A_2 & E & E \\ \cdot & T_2 & T_2 \\ T_2 & \cdot & \cdot \end{bmatrix} = \frac{1}{6\sqrt{2}}$	$\begin{bmatrix} A_2 & \cdot & T_2 \\ \cdot & E & \cdot \\ T_2 & T_2 & E \end{bmatrix} = 0$	$\begin{bmatrix} A_2 & \cdot & T_2 \\ T_2 & E & \cdot \\ \cdot & T_2 & E \end{bmatrix} = \frac{1}{9}$
123	$\begin{bmatrix} A_2 & \cdot & T_2 \\ T_2 & E & T_2 \\ \cdot & \cdot & E \end{bmatrix} = \frac{1}{9}$	$\begin{bmatrix} A_2 & E & E \\ \cdot & \cdot & T_2 \\ T_2 & T_2 & T_2 \end{bmatrix} = -\frac{1}{6\sqrt{6}}$	$\begin{bmatrix} A_2 & E & E \\ \cdot & T_2 & T_2 \\ T_2 & \cdot & T_2 \end{bmatrix} = \frac{1}{6\sqrt{6}}$
124	$\begin{bmatrix} A_2 & \cdot & T_2 \\ \cdot & E & T_2 \\ T_2 & T_2 & E \end{bmatrix} = \frac{1}{9}$	$\begin{bmatrix} A_2 & \cdot & T_2 \\ T_2 & E & T_2 \\ \cdot & T_2 & E \end{bmatrix} = 0$	$\begin{bmatrix} A_2 & E & E \\ \cdot & T_2 & T_2 \\ T_2 & T_2 & T_2 \end{bmatrix} = \frac{1}{6\sqrt{2}}$
140	$\begin{bmatrix} A_2 & E & E \\ E & \cdot & \cdot \\ E & \cdot & \cdot \end{bmatrix} = 0$	$\begin{bmatrix} A_2 & E & E \\ E & \cdot & \cdot \\ E & \cdot & T_2 \end{bmatrix} = \frac{1}{6}$	$\begin{bmatrix} A_2 & E & E \\ \cdot & E & \cdot \\ T_2 & E & \cdot \end{bmatrix} = -\frac{1}{6\sqrt{2}}$
142	$\begin{bmatrix} A_2 & E & E \\ E & \cdot & \cdot \\ E & T_2 & T_2 \end{bmatrix} = 0$	$\begin{bmatrix} A_2 & E & E \\ E & \cdot & T_2 \\ E & T_2 & \cdot \end{bmatrix} = 0$	$\begin{bmatrix} A_2 & E & E \\ \cdot & E & \cdot \\ T_2 & E & T_2 \end{bmatrix} = -\frac{1}{6\sqrt{2}}$
142	$\begin{bmatrix} A_2 & E & E \\ \cdot & E & T_2 \\ T_2 & E & \cdot \end{bmatrix} = \frac{1}{6\sqrt{2}}$	$\begin{bmatrix} A_2 & E & E \\ E & \cdot & T_2 \\ E & T_2 & T_2 \end{bmatrix} = \frac{1}{6}$	$\begin{bmatrix} A_2 & E & E \\ \cdot & E & T_2 \\ T_2 & E & T_2 \end{bmatrix} = \frac{1}{6\sqrt{2}}$
144	$\begin{bmatrix} A_2 & E & E \\ E & T_2 & T_2 \\ E & T_2 & T_2 \end{bmatrix} = 0$	$\begin{bmatrix} A_2 & E & E \\ E & E & E \\ E & E & E \end{bmatrix} = 0$	$\begin{bmatrix} A_2 & E & E \\ \cdot & E & \cdot \\ T_2 & A_2 & \cdot \end{bmatrix} = \frac{1}{6}$
232	$\begin{bmatrix} A_2 & E & E \\ \cdot & E & T_2 \\ T_2 & A_2 & \cdot \end{bmatrix} = \frac{1}{6}$	$\begin{bmatrix} A_2 & E & E \\ T_2 & E & \cdot \\ \cdot & A_2 & T_2 \end{bmatrix} = \frac{1}{6}$	$\begin{bmatrix} A_2 & E & E \\ T_2 & E & T_2 \\ \cdot & A_2 & T_2 \end{bmatrix} = \frac{1}{6}$
270	$\begin{bmatrix} A_2 & E & E \\ E & A_2 & E \\ E & E & E \end{bmatrix} = \frac{1}{4}$	$\begin{bmatrix} A_2 & E & E \\ E & A_2 & E \\ E & E & A_2 \end{bmatrix} = -\frac{1}{4}$	

APPENDIX D

Tables for the Dihedral Groups

D1 THE GROUP D_2

D_2 is commutative, so the concept of component is superfluous, as discussed in Section 3.1. However, we could write $V(abc) = \delta(a, b, c)$,

$$W\begin{pmatrix} a & b & c \\ d & e & f \end{pmatrix} = V(abc)V(aef)V(bfd)V(cde)$$

and analogously for X, if we wished.

D2 THE GROUP D_3

D_3, referred to the (111) axis, may be regarded as a subgroup of the octahedral group O. Functions belonging to the irreducible representations A_1, A_2, and E of O then become functions belonging to the irreducible representations A_1, A_2, and E, respectively, of D_3. The components (θ, ϵ) of E for O give rise, respectively, to (c, s) for E for D_3, and it follows from comparison of Table C2.1 with Tables 3.1 and 3.2 that the phases of the V coefficients have been chosen to correspond in the two groups. Therefore, any W or X coefficient for D_3 may be obtained by looking it up in Tables C3.1 or C3.2.

D3 THE GROUP D_4

Table D3.1. Direct products of one-valued representations

	A_1	A_2	B_1	B_2	E
A_1	A_1	A_2	B_1	B_2	E
A_2	A_2	A_1	B_2	B_1	E
B_1	B_1	B_2	A_1	A_2	E
B_2	B_2	B_1	A_2	A_1	E
E	E	E	E	E	$A_1 + A_2 + B_1 + B_2$

Table D3.2. V coefficients. A dot represents zero

$(-1)^a = -1$ for $a = A_2$, $\quad (-1)^a = 1$ otherwise

$[-1]^a = 1$ always

$$V\begin{pmatrix} a & a & A_1 \\ \alpha & \beta & \iota \end{pmatrix} = \lambda(a)^{-1/2}\delta_{\alpha\beta}$$

$$V\begin{pmatrix} A_2 & B_1 & B_2 \\ \iota & \iota & \iota \end{pmatrix} = 1, \qquad V\begin{pmatrix} A_2 & B_2 & B_1 \\ \iota & \iota & \iota \end{pmatrix} = -1$$

		V			
E	E	A_1	A_2	B_1	B_2
c	c	$\frac{1}{\sqrt{2}}$	.	$-\frac{1}{\sqrt{2}}$	.
s	s	$\frac{1}{\sqrt{2}}$	.	$\frac{1}{\sqrt{2}}$	.
c	s	.	$\frac{1}{\sqrt{2}}$	.	$\frac{1}{\sqrt{2}}$
s	c	.	$-\frac{1}{\sqrt{2}}$	.	$\frac{1}{\sqrt{2}}$

		V			
E	E	A_1	A_2	B_1	B_2
1	1	.	.	$\frac{1}{\sqrt{2}}$	$-\frac{i}{\sqrt{2}}$
1	-1	$\frac{1}{\sqrt{2}}$	$-\frac{i}{\sqrt{2}}$	.	.
-1	1	$\frac{1}{\sqrt{2}}$	$\frac{i}{\sqrt{2}}$	.	.
-1	-1	.	.	$\frac{1}{\sqrt{2}}$	$\frac{i}{\sqrt{2}}$

Table D3.3. Table of W. Any W not containing A_1 may be rearranged into one of these by using the symmetry operations of Section 4.2. The symbol W is omitted.

$$\begin{pmatrix} A_2 & B_1 & B_2 \\ A_2 & B_1 & B_2 \end{pmatrix} = 1 \qquad \begin{pmatrix} A_2 & E & E \\ B_2 & E & E \end{pmatrix} = \tfrac{1}{2}$$

$$\begin{pmatrix} A_2 & B_1 & B_2 \\ E & E & E \end{pmatrix} = \frac{1}{\sqrt{2}} \qquad \begin{pmatrix} B_1 & E & E \\ B_1 & E & E \end{pmatrix} = \tfrac{1}{2}$$

$$\begin{pmatrix} A_2 & E & E \\ A_2 & E & E \end{pmatrix} = \tfrac{1}{2} \qquad \begin{pmatrix} B_1 & E & E \\ B_2 & E & E \end{pmatrix} = -\tfrac{1}{2}$$

$$\begin{pmatrix} A_2 & E & E \\ B_1 & E & E \end{pmatrix} = \tfrac{1}{2} \qquad \begin{pmatrix} B_2 & E & E \\ B_2 & E & E \end{pmatrix} = \tfrac{1}{2}$$

Simple formulae for the X are given in Section 8.3.

D4 THE GROUP D_5

Table D4.1. Direct products of one-valued representations.

	A_1	A_2	E_1	E_2
A_1	A_1	A_2	E_1	E_2
A_2	A_2	A_1	E_1	E_2
E_1	E_1	E_1	$A_1 + A_2 + E_2$	$E_1 + E_2$
E_2	E_2	E_2	$E_1 + E_2$	$A_1 + A_2 + E_1$

Table D4.2. V coefficients. A dot represents zero.

$$(-1)^a = -1 \text{ for } a = A_2, \qquad (-1)^a = 1 \text{ otherwise}$$
$$[-1]^a = 1 \text{ always}$$

$$V\begin{pmatrix} a & a & A_1 \\ \alpha & \beta & \iota \end{pmatrix} = \lambda(a)^{-1/2}\delta_{\alpha\beta}$$

E_1	E_1	A_1	A_2	E_2 c	E_2 s
c	c	$\dfrac{1}{\sqrt{2}}$	.	$\frac{1}{2}$	.
s	s	$\dfrac{1}{\sqrt{2}}$	.	$-\frac{1}{2}$	.
c	s	.	$\dfrac{1}{\sqrt{2}}$	.	$\frac{1}{2}$
s	c	.	$-\dfrac{1}{\sqrt{2}}$	.	$\frac{1}{2}$

E_2	E_2	A_1	A_2	E_1 c	E_1 s
c	c	$\dfrac{1}{\sqrt{2}}$	.	$-\frac{1}{2}$	.
s	s	$\dfrac{1}{\sqrt{2}}$	.	$\frac{1}{2}$	.
c	s	.	$\dfrac{1}{\sqrt{2}}$	.	$\frac{1}{2}$
s	c	.	$-\dfrac{1}{\sqrt{2}}$	.	$\frac{1}{2}$

E_1	E_1	A_1	A_2	E_2 1	-1
1	1	.	.	.	$-\dfrac{i}{\sqrt{2}}$
1	-1	$\dfrac{1}{\sqrt{2}}$	$-\dfrac{i}{\sqrt{2}}$	.	.
-1	1	$\dfrac{1}{\sqrt{2}}$	$\dfrac{i}{\sqrt{2}}$	.	.
-1	-1	.	.	$\dfrac{i}{\sqrt{2}}$	.

E_2	E_2	A_1	A_2	E_1 1	-1
1	1	.	.	$-\dfrac{i}{\sqrt{2}}$	.
1	-1	$\dfrac{1}{\sqrt{2}}$	$-\dfrac{i}{\sqrt{2}}$	.	.
-1	1	$\dfrac{1}{\sqrt{2}}$	$\dfrac{i}{\sqrt{2}}$	.	.
-1	-1	.	.	.	$\dfrac{i}{\sqrt{2}}$

Table D4.3. Table of W. The symbol W is omitted.

$$\begin{pmatrix} A_2 & E_1 & E_1 \\ A_2 & E_1 & E_1 \end{pmatrix} = \begin{pmatrix} A_2 & E_2 & E_2 \\ A_2 & E_2 & E_2 \end{pmatrix} = \tfrac{1}{2} \qquad \begin{pmatrix} A_2 & E_1 & E_1 \\ E_1 & E_2 & E_2 \end{pmatrix} = -\tfrac{1}{2}$$

$$\begin{pmatrix} A_2 & E_1 & E_1 \\ E_2 & E_1 & E_1 \end{pmatrix} = \begin{pmatrix} A_2 & E_1 & E_1 \\ E_2 & E_2 & E_2 \end{pmatrix} = \tfrac{1}{2} \qquad \begin{pmatrix} A_2 & E_2 & E_2 \\ E_1 & E_2 & E_2 \end{pmatrix} = \tfrac{1}{2}$$

$$\begin{pmatrix} E_1 & E_1 & E_2 \\ E_1 & E_1 & E_2 \end{pmatrix} = \begin{pmatrix} E_1 & E_2 & E_2 \\ E_1 & E_2 & E_2 \end{pmatrix} = 0 \qquad \begin{pmatrix} E_1 & E_1 & E_2 \\ E_2 & E_1 & E_2 \end{pmatrix} = \tfrac{1}{2}$$

Table D4.4. Table of X. The symbol X is omitted, and A_2, E_1, E_2 are replaced with ., 1, 2, respectively. The X are partially ordered in terms of increasing number of A_2 representations.

$$\begin{bmatrix} 1 & 2 & 2 \\ 2 & 1 & 2 \\ 2 & 2 & 1 \end{bmatrix} = \begin{bmatrix} 1 & 1 & 2 \\ 2 & 2 & 1 \\ 2 & 2 & 1 \end{bmatrix} = \begin{bmatrix} 1 & 1 & 2 \\ 1 & 1 & 2 \\ 2 & 2 & 1 \end{bmatrix} = \begin{bmatrix} 1 & 1 & 2 \\ 1 & 2 & 1 \\ 2 & 1 & 1 \end{bmatrix} = \tfrac{1}{4}$$

$$\begin{bmatrix} 1 & 1 & 2 \\ 1 & 2 & 2 \\ 2 & 1 & 1 \end{bmatrix} = \begin{bmatrix} 1 & 1 & 2 \\ 1 & 2 & 2 \\ 2 & 2 & 1 \end{bmatrix} = \begin{bmatrix} 1 & 1 & 2 \\ 1 & 2 & 1 \\ 2 & 1 & 2 \end{bmatrix} = 0$$

$$\begin{bmatrix} . & 1 & 1 \\ 1 & 1 & 2 \\ 1 & 2 & 1 \end{bmatrix} = \begin{bmatrix} . & 1 & 1 \\ 1 & 2 & 2 \\ 1 & 2 & 2 \end{bmatrix} = \begin{bmatrix} . & 2 & 2 \\ 2 & 2 & 1 \\ 2 & 1 & 2 \end{bmatrix} = \begin{bmatrix} . & 2 & 2 \\ 2 & 1 & 1 \\ 2 & 1 & 1 \end{bmatrix} = 0$$

$$\begin{bmatrix} . & 1 & 1 \\ 1 & 1 & 2 \\ 1 & 2 & 2 \end{bmatrix} = \begin{bmatrix} . & 1 & 1 \\ 2 & 1 & 2 \\ 2 & 2 & 1 \end{bmatrix} = \begin{bmatrix} . & 2 & 2 \\ 2 & 2 & 1 \\ 2 & 1 & 1 \end{bmatrix} = \tfrac{1}{4}$$

$$\begin{bmatrix} . & 1 & 1 \\ 1 & . & 1 \\ 1 & 1 & 2 \end{bmatrix} = \begin{bmatrix} . & 1 & 1 \\ 2 & . & 2 \\ 2 & 1 & 1 \end{bmatrix} = \begin{bmatrix} . & 1 & 1 \\ 2 & . & 2 \\ 2 & 1 & 2 \end{bmatrix} = \begin{bmatrix} . & 2 & 2 \\ 2 & . & 2 \\ 2 & 2 & 1 \end{bmatrix} = \tfrac{1}{4}$$

$$\begin{bmatrix} . & 1 & 1 \\ 1 & . & 1 \\ 1 & 1 & . \end{bmatrix} = \begin{bmatrix} . & 2 & 2 \\ 2 & . & 2 \\ 2 & 2 & . \end{bmatrix} = -\tfrac{1}{4}$$

D5 THE GROUP D_6

D_6 is the direct product of D_3 and C_2 and, therefore, its V, W, and X coefficients may be deduced at once from those of D_3 and C_2 according to Eqs. (3.6) and (3.8) and the analogous formula for X. If we write A for the unit representation of C_2 and B for the other, the representations of $D_3 \times C_2$ correspond to those of D_6, in the usual notation for the latter, according to the following table:

D_3	A_1	A_2	E	A_1	A_2	E
C_2	A	A	A	B	B	B
D_6	A_1	A_2	E_2	B_1	B_2	E_1

We now consider how to obtain W. Precisely analogous remarks apply to X. For C_2 we should naturally define $V(abc) = \delta(a, b, c)$ as we did under $D1$. Then $W = 1$ or 0 as

$$\delta(a, b, c)\delta(a, e, f)\delta(b, f, d)\delta(c, d, e) = 1 \text{ or } 0$$

Hence, when the corresponding quantity derived from a W for D_6 is equal to 1, we can deduce from Eq. (3.8) that

$$W\begin{pmatrix} aa' & bb' & cc' \\ dd' & ee' & ff' \end{pmatrix} = W\begin{pmatrix} a & b & c \\ d & e & f \end{pmatrix}$$

Here the unprimed symbols refer to D_3 and the primed ones to C_2. However, there is a further complication, in that the choice of phases for components of representations and for V coefficients are not the same for $D_3 \times C_2$ as for D_6 when the definitions of Sections 3.2 and 3.4 are used. To obtain the correct phases for W coefficients for D_6, we must multiply those of $D_3 \times C_2$ by $(-1)^{m+n}$, where m = number of times any of the ordered trios $A_2B_2B_1$, $B_2B_1A_2$, $B_1A_2B_2$, $E_1E_2B_2$, $E_2B_2E_1$, $B_2E_1E_2$ occur among the constituent V coefficients in W, and n is the number of times the unordered trios $E_2^2A_2$, $E_1^2E_2$, $E_1E_2B_1$ occur. For example,

$$W\begin{pmatrix} A_2 & E_1 & E_1 \\ B_2 & E_2 & E_2 \end{pmatrix} = W\begin{pmatrix} A_2 & E & E \\ A_2 & E & E \end{pmatrix} = \tfrac{1}{2}$$

because $m = n = 1$. It is now easy to deduce all W and X. Because of the importance of the group D_6, the W are tabulated below.

Table D5.1. Table of W. All W with six constituent E representations are zero. The symbol W is omitted.

$$\begin{pmatrix} A_2 & B_1 & B_2 \\ A_2 & B_1 & B_2 \end{pmatrix} = 1, \qquad \begin{pmatrix} A_2 & B_1 & B_2 \\ E_2 & E_1 & E_1 \end{pmatrix} = \begin{pmatrix} A_2 & B_1 & B_2 \\ E_1 & E_2 & E_2 \end{pmatrix} = \frac{1}{\sqrt{2}}$$

$$\begin{pmatrix} A_2 & E_1 & E_1 \\ A_2 & E_1 & E_1 \end{pmatrix} = \begin{pmatrix} A_2 & E_2 & E_2 \\ A_2 & E_2 & E_2 \end{pmatrix} = \begin{pmatrix} A_2 & E_1 & E_1 \\ B_1 & E_2 & E_2 \end{pmatrix} = \begin{pmatrix} A_2 & E_1 & E_1 \\ B_2 & E_2 & E_2 \end{pmatrix} = \tfrac{1}{2}$$

$$\begin{pmatrix} B_1 & E_1 & E_2 \\ B_1 & E_1 & E_2 \end{pmatrix} = \begin{pmatrix} B_2 & E_1 & E_2 \\ B_2 & E_1 & E_2 \end{pmatrix} = \tfrac{1}{2}, \qquad \begin{pmatrix} B_1 & E_1 & E_2 \\ B_2 & E_1 & E_2 \end{pmatrix} = -\tfrac{1}{2}$$

D6 THE GROUP D_∞

Table D6.1. Direct products of one-valued representations.

	A_1	A_2	E_n
A_1	A_1	A_2	E_n
A_2	A_2	A_1	E_n
E_n	E_n	E_n	$A_1 + A_2 + E_{2n}$

$E_m E_n = E_{|m-n|} + E_{m+n}$
when $m \neq n$

Table D6.2. Non-zero V coefficients. A dot represents zero.

$(-1)^a = -1$ for $a = A_2$, $\quad (-1)^a = 1$ otherwise
$[-1]^a = 1$ always

$$V\begin{pmatrix} a & a & A_1 \\ \alpha & \beta & \iota \end{pmatrix} = \lambda(a)^{-1/2}\delta_{\alpha\beta}$$

E_m	E_n	E_{m+n}	V
c	c	c	$\frac{1}{2}$
s	s	c	$-\frac{1}{2}$
s	c	s	$\frac{1}{2}$
c	s	s	$\frac{1}{2}$
1	1	-1	$-i/\sqrt{2}$
-1	-1	1	$i/\sqrt{2}$

E_n	E_n	A_1	A_2	E_{2n} c	E_{2n} s
c	c	$\frac{1}{\sqrt{2}}$	·	$\frac{1}{2}$	·
s	s	$\frac{1}{\sqrt{2}}$	·	$-\frac{1}{2}$	·
c	s	·	$\frac{1}{\sqrt{2}}$	·	$\frac{1}{2}$
s	c	·	$-\frac{1}{\sqrt{2}}$	·	$\frac{1}{2}$
E_n	E_n	A_1	A_2	1	-1
1	1	·	·	·	$-\frac{i}{\sqrt{2}}$
1	-1	$\frac{1}{\sqrt{2}}$	$-\frac{i}{\sqrt{2}}$	·	·
-1	1	$\frac{1}{\sqrt{2}}$	$\frac{i}{\sqrt{2}}$	·	·
-1	-1	·	·	$\frac{i}{\sqrt{2}}$	·

Table D6.3. Table of W. The symbol W is omitted.

$$\begin{pmatrix} A_2 & E_n & E_n \\ A_2 & E_n & E_n \end{pmatrix} = \tfrac{1}{2}, \qquad \begin{pmatrix} E_m & E_n & E_p \\ E_m & E_n & E_p \end{pmatrix} = 0$$

$$\begin{pmatrix} A_2 & E_m & E_m \\ E_p & E_n & E_n \end{pmatrix} = \tfrac{1}{2} \qquad \text{when } p = m + n$$

$$= -\tfrac{1}{2} \qquad \text{when } n = m + p \quad \text{or} \quad m = n + p$$

$$\begin{pmatrix} E_m & E_n & E_p \\ E_q & E_r & E_s \end{pmatrix} = \tfrac{1}{2} \qquad \text{when } m = n + p = r + s = p + q + s$$

APPENDIX E

Values of the Coefficients Ω for Spin-Orbit Coupling

In this Appendix are tabulated values of the coefficients Ω defined in Eq. (9.31) for the octahedral group, with spins S, $S' \leqslant 2$. The phases adopted for the coupling coefficients are identical with those used in my previous book (26), except for the $\langle abc\gamma \mid ab\alpha\beta\rangle$ with a, b, c one-valued representations of O. These have their phases redefined consistently with the V coefficients given in Appendix C.

In reference (26) we coupled spin with space for a term ${}^{2S+1}T_1$ by pretending it was a ${}^{2S+1}P$ atomic term and using the Wigner coefficients $\langle S1JM \mid S1M_SM_L\rangle$ to give the states of ${}^{2S+1}T_{1J}$ $(J = S, S \pm 1)$. It follows rather easily from this and the selection rule $\Delta J = 0$ for matrix elements of spin-orbit coupling between atomic levels that $\Omega = 0$ for any representation t occurring in ${}^{2S+1}A_1$ and ${}^{2S'+1}T_{1J}$ unless $J = S$. Similarly, when $J = S$, it follows from Eqs. (9.13) and (9.31) that for corresponding representations t

$$\Omega_{SS}\begin{pmatrix} S & S' & T_1 \\ T_1 & A_1 & t \end{pmatrix} = [3(2S+1)]^{-1/2} \tag{E1}$$

The phases of ${}^{2S+1}A_2$ and ${}^{2S+1}T_2$ can be chosen so that Eq. (E1) also holds for them:

$$\Omega_{SS}\begin{pmatrix} S & S' & T_1 \\ T_2 & A_2 & t \end{pmatrix} = [3(2S+1)]^{-1/2} \tag{E2}$$

and again $\Omega = 0$ unless $J = S$. Because of the simplicity of these formulae, Ω is only tabulated for h, $h' = E$, T_1 or T_2.

Other general formulae can be given relating certain Ω to $\overline{W}$ coefficients. The simplest of these are

$$\Omega_{JJ'}\begin{pmatrix} S & S' & T_1 \\ T_1 & T_1 & c \end{pmatrix} = \Omega_{JJ'}\begin{pmatrix} S & S' & T_1 \\ T_2 & T_2 & c \end{pmatrix}$$
$$= (-1)^{J+S}\delta_{JJ'}\overline{W}\begin{pmatrix} 1 & 1 & 1 \\ S & S' & J \end{pmatrix} \tag{E3}$$

Table E1. Values of $\Omega_{JJ'}\begin{pmatrix} S & S' & T_1 \\ h' & h & t \end{pmatrix}$. The symbol ${}^{2S+1}h_J$ is written at the left and ${}^{2S'+1}h'_{J'}$ along the top of each table. J is omitted when it is redundant.

$t = A_1$	${}^3T_1$	${}^5E$	${}^5T_2$
${}^3T_1$	$\frac{1}{3}$	$\frac{1}{\sqrt{15}}$	$-\frac{1}{\sqrt{15}}$
${}^5E$	$\frac{1}{\sqrt{15}}$	0	$-\frac{1}{\sqrt{15}}$
${}^5T_2$	$-\frac{1}{\sqrt{15}}$	$-\frac{1}{\sqrt{15}}$	$-\frac{1}{3\sqrt{5}}$

$t = A_2$	${}^3T_2$	${}^5E$	${}^5T_1$
${}^3T_2$	$\frac{1}{3}$	$-\frac{1}{\sqrt{15}}$	$-\frac{1}{\sqrt{15}}$
${}^5E$	$\frac{1}{\sqrt{15}}$	0	$\frac{1}{\sqrt{15}}$
${}^5T_1$	$-\frac{1}{\sqrt{15}}$	$-\frac{1}{\sqrt{15}}$	$-\frac{1}{3\sqrt{5}}$

$t = E$	${}^1E$	${}^3T_1$	${}^3T_2$	${}^5E$	${}^5T_1$	${}^5T_2$
${}^1E$	0	$\frac{1}{\sqrt{6}}$	$-\frac{1}{\sqrt{6}}$	0	0	0
${}^3T_1$	$\frac{1}{\sqrt{6}}$	$-\frac{1}{6}$	$\frac{1}{2\sqrt{3}}$	$\frac{1}{\sqrt{30}}$	$\frac{1}{2\sqrt{5}}$	$-\frac{1}{2\sqrt{15}}$
${}^3T_2$	$\frac{1}{\sqrt{6}}$	$-\frac{1}{2\sqrt{3}}$	$-\frac{1}{6}$	$-\frac{1}{\sqrt{30}}$	$-\frac{1}{2\sqrt{15}}$	$-\frac{1}{2\sqrt{5}}$
${}^5E$	0	$\frac{1}{\sqrt{30}}$	$\frac{1}{\sqrt{30}}$	0	$\frac{1}{\sqrt{30}}$	$-\frac{1}{\sqrt{30}}$
${}^5T_1$	0	$-\frac{1}{2\sqrt{5}}$	$-\frac{1}{2\sqrt{15}}$	$-\frac{1}{\sqrt{30}}$	$\frac{1}{6\sqrt{5}}$	$-\frac{1}{2\sqrt{15}}$
${}^5T_2$	0	$-\frac{1}{2\sqrt{15}}$	$\frac{1}{2\sqrt{5}}$	$-\frac{1}{\sqrt{30}}$	$\frac{1}{2\sqrt{15}}$	$\frac{1}{6\sqrt{5}}$

$t = T_1$	${}^1T_1$	${}^3E$	${}^3T_1$	${}^3T_2$	${}^5E$	${}^5T_{11}$	${}^5T_{13}$	${}^5T_{22}$	${}^5T_{23}$
${}^1T_1$	0	$\frac{1}{3}$	$-\frac{1}{3}$	$\frac{1}{3}$	0	0	0	0	0
${}^3E$	$\frac{1}{3}$	0	$-\frac{1}{6}$	$\frac{1}{2\sqrt{3}}$	0	$\frac{1}{30}$	$-\frac{\sqrt{3}}{5\sqrt{2}}$	$-\frac{1}{2\sqrt{15}}$	$\frac{1}{\sqrt{30}}$
${}^3T_1$	$\frac{1}{3}$	$\frac{1}{6}$	$\frac{1}{6}$	$\frac{1}{6}$	$\frac{1}{2\sqrt{5}}$	$-\frac{1}{6}$	0	$-\frac{1}{6\sqrt{5}}$	$-\frac{\sqrt{2}}{3\sqrt{5}}$
${}^3T_2$	$\frac{1}{3}$	$\frac{1}{2\sqrt{3}}$	$-\frac{1}{6}$	$-\frac{1}{6}$	$-\frac{1}{2\sqrt{15}}$	$\frac{1}{30}$	$\frac{\sqrt{2}}{5\sqrt{3}}$	$-\frac{1}{2\sqrt{5}}$	0
${}^5E$	0	0	$\frac{1}{2\sqrt{5}}$	$\frac{1}{2\sqrt{15}}$	0	$-\frac{1}{10}$	$-\frac{1}{5\sqrt{6}}$	$-\frac{1}{2\sqrt{15}}$	$-\frac{1}{\sqrt{30}}$
${}^5T_{11}$	0	$\frac{1}{30}$	$\frac{1}{6}$	$\frac{1}{30}$	$\frac{1}{10}$	$\frac{1}{2\sqrt{5}}$	0	$-\frac{1}{30}$	$-\frac{\sqrt{2}}{15}$
${}^5T_{13}$	0	$-\frac{\sqrt{3}}{5\sqrt{2}}$	0	$\frac{\sqrt{2}}{5\sqrt{3}}$	$\frac{1}{5\sqrt{6}}$	0	$-\frac{1}{3\sqrt{5}}$	$-\frac{\sqrt{2}}{5\sqrt{3}}$	$\frac{1}{5\sqrt{3}}$
${}^5T_{22}$	0	$\frac{1}{2\sqrt{15}}$	$-\frac{1}{6\sqrt{5}}$	$\frac{1}{2\sqrt{5}}$	$-\frac{1}{2\sqrt{15}}$	$\frac{1}{30}$	$\frac{\sqrt{2}}{5\sqrt{3}}$	$\frac{1}{6\sqrt{5}}$	0
${}^5T_{23}$	0	$-\frac{1}{\sqrt{30}}$	$-\frac{\sqrt{2}}{3\sqrt{5}}$	0	$-\frac{1}{\sqrt{30}}$	$\frac{\sqrt{2}}{15}$	$-\frac{1}{5\sqrt{3}}$	0	$-\frac{1}{3\sqrt{5}}$

Table E1. (Continued)

$t = T_2$	$^1T_2$	^3E	$^3T_1$	$^3T_2$	^5E	$^5T_{12}$	$^5T_{13}$	$^5T_{21}$	$^5T_{23}$
$^1T_2$	0	$-\frac{1}{3}$	$\frac{1}{3}$	$-\frac{1}{3}$	0	0	0	0	0
^3E	$\frac{1}{3}$	0	$-\frac{1}{2\sqrt{3}}$	$-\frac{1}{6}$	0	$\frac{1}{2\sqrt{15}}$	$-\frac{1}{\sqrt{30}}$	$\frac{1}{30}$	$-\frac{\sqrt{3}}{5\sqrt{2}}$
$^3T_1$	$-\frac{1}{3}$	$\frac{1}{2\sqrt{3}}$	$-\frac{1}{6}$	$-\frac{1}{6}$	$-\frac{1}{2\sqrt{15}}$	$-\frac{1}{2\sqrt{5}}$	0	$\frac{1}{30}$	$\frac{\sqrt{2}}{5\sqrt{3}}$
$^3T_2$	$\frac{1}{3}$	$-\frac{1}{6}$	$\frac{1}{6}$	$\frac{1}{6}$	$-\frac{1}{2\sqrt{5}}$	$-\frac{1}{6\sqrt{5}}$	$-\frac{\sqrt{2}}{3\sqrt{5}}$	$-\frac{1}{6}$	0
^5E	0	0	$-\frac{1}{2\sqrt{15}}$	$\frac{1}{2\sqrt{5}}$	0	$\frac{1}{2\sqrt{15}}$	$\frac{1}{\sqrt{30}}$	$-\frac{1}{10}$	$-\frac{1}{5\sqrt{6}}$
$^5T_{12}$	0	$\frac{1}{2\sqrt{15}}$	$\frac{1}{2\sqrt{5}}$	$-\frac{1}{6\sqrt{5}}$	$-\frac{1}{2\sqrt{15}}$	$\frac{1}{6\sqrt{5}}$	0	$\frac{1}{30}$	$\frac{\sqrt{2}}{5\sqrt{3}}$
$^5T_{13}$	0	$-\frac{1}{\sqrt{30}}$	0	$-\frac{\sqrt{2}}{3\sqrt{5}}$	$-\frac{1}{\sqrt{30}}$	0	$-\frac{1}{3\sqrt{5}}$	$\frac{\sqrt{2}}{15}$	$-\frac{1}{5\sqrt{3}}$
$^5T_{21}$	0	$-\frac{1}{30}$	$\frac{1}{30}$	$\frac{1}{6}$	$-\frac{1}{10}$	$-\frac{1}{30}$	$-\frac{\sqrt{2}}{15}$	$\frac{1}{2\sqrt{5}}$	0
$^5T_{23}$	0	$\frac{\sqrt{3}}{5\sqrt{2}}$	$\frac{\sqrt{2}}{5\sqrt{3}}$	0	$-\frac{1}{5\sqrt{6}}$	$-\frac{\sqrt{2}}{5\sqrt{3}}$	$\frac{1}{5\sqrt{3}}$	0	$-\frac{1}{3\sqrt{5}}$

$t = E'$	$^2T_1$	^4E	$^4T_1$	$^4T_2$
$^2T_1$	$\frac{1}{3}$	$-\frac{1}{2\sqrt{3}}$	$-\frac{1}{6}$	$\frac{1}{2\sqrt{3}}$
^4E	$-\frac{1}{2\sqrt{3}}$	0	$\frac{1}{2\sqrt{30}}$	$-\frac{\sqrt{3}}{2\sqrt{10}}$
$^4T_1$	$\frac{1}{6}$	$-\frac{1}{2\sqrt{30}}$	$\frac{\sqrt{5}}{6\sqrt{2}}$	$\frac{1}{2\sqrt{30}}$
$^4T_2$	$\frac{1}{2\sqrt{3}}$	$-\frac{\sqrt{3}}{2\sqrt{10}}$	$-\frac{1}{2\sqrt{30}}$	$-\frac{1}{2\sqrt{10}}$

$t = E''$	$^2T_2$	^4E	$^4T_1$	$^4T_2$
$^2T_2$	$\frac{1}{3}$	$\frac{1}{2\sqrt{3}}$	$\frac{1}{2\sqrt{3}}$	$-\frac{1}{6}$
^4E	$-\frac{1}{2\sqrt{3}}$	0	$\frac{\sqrt{3}}{2\sqrt{10}}$	$\frac{1}{2\sqrt{30}}$
$^4T_1$	$\frac{1}{2\sqrt{3}}$	$-\frac{\sqrt{3}}{2\sqrt{10}}$	$-\frac{1}{2\sqrt{10}}$	$-\frac{1}{2\sqrt{30}}$
$^4T_2$	$\frac{1}{6}$	$\frac{1}{2\sqrt{30}}$	$\frac{1}{2\sqrt{30}}$	$\frac{\sqrt{5}}{6\sqrt{2}}$

Table E1. (Continued)

$t = U'$	^2E	$^2T_1$	$^2T_2$	^4E	$^4T_{1\ 3/2}$	$^4T_{1\ 5/2}$	$^4T_{2\ 3/2}$	$^4T_{2\ 5/2}$
^2E	0	$\frac{1}{2\sqrt{3}}$	$-\frac{1}{2\sqrt{3}}$	0	$-\frac{1}{2\sqrt{30}}$	$-\frac{\sqrt{3}}{2\sqrt{10}}$	$\frac{1}{2\sqrt{30}}$	$\frac{\sqrt{3}}{2\sqrt{10}}$
$^2T_1$	$-\frac{1}{2\sqrt{3}}$	$-\frac{1}{6}$	$\frac{1}{2\sqrt{3}}$	$\frac{1}{2\sqrt{6}}$	$-\frac{\sqrt{5}}{6\sqrt{2}}$	0	$-\frac{1}{2\sqrt{30}}$	$\frac{1}{\sqrt{30}}$
$^2T_2$	$-\frac{1}{2\sqrt{3}}$	$-\frac{1}{2\sqrt{3}}$	$-\frac{1}{6}$	$-\frac{1}{2\sqrt{6}}$	$\frac{1}{2\sqrt{30}}$	$-\frac{1}{\sqrt{30}}$	$-\frac{\sqrt{5}}{6\sqrt{2}}$	0
^4E	0	$\frac{1}{2\sqrt{6}}$	$\frac{1}{2\sqrt{6}}$	0	$-\frac{\sqrt{2}}{5\sqrt{3}}$	$\frac{\sqrt{3}}{10\sqrt{2}}$	$-\frac{\sqrt{2}}{5\sqrt{3}}$	$\frac{\sqrt{3}}{10\sqrt{2}}$
$^4T_{1\ 3/2}$	$-\frac{1}{2\sqrt{30}}$	$\frac{\sqrt{5}}{6\sqrt{2}}$	$\frac{1}{2\sqrt{30}}$	$\frac{\sqrt{2}}{5\sqrt{3}}$	$\frac{1}{3\sqrt{10}}$	0	$-\frac{\sqrt{2}}{5\sqrt{15}}$	$\frac{2\sqrt{2}}{5\sqrt{15}}$
$^4T_{1\ 5/2}$	$-\frac{\sqrt{3}}{2\sqrt{10}}$	0	$-\frac{1}{\sqrt{30}}$	$-\frac{\sqrt{3}}{10\sqrt{2}}$	0	$-\frac{1}{2\sqrt{10}}$	$\frac{2\sqrt{2}}{5\sqrt{15}}$	$\frac{3\sqrt{3}}{10\sqrt{10}}$
$^4T_{2\ 3/2}$	$-\frac{1}{2\sqrt{30}}$	$-\frac{1}{2\sqrt{30}}$	$\frac{\sqrt{5}}{6\sqrt{2}}$	$-\frac{\sqrt{2}}{5\sqrt{3}}$	$\frac{\sqrt{2}}{5\sqrt{15}}$	$-\frac{2\sqrt{2}}{5\sqrt{15}}$	$\frac{1}{3\sqrt{10}}$	0
$^4T_{2\ 5/2}$	$-\frac{\sqrt{3}}{2\sqrt{10}}$	$\frac{1}{\sqrt{30}}$	0	$\frac{\sqrt{3}}{10\sqrt{2}}$	$-\frac{2\sqrt{2}}{5\sqrt{15}}$	$-\frac{3\sqrt{3}}{10\sqrt{10}}$	0	$-\frac{1}{2\sqrt{10}}$

APPENDIX F

Acknowledgements and References

The work described in the present book has various foundations in the published literature. The first and most important is in the theory of the irreducible tensor method for systems classified under the full rotation group, including spin rotations [refs. (1)–(5)]. The second is the conceptual and mathematical apparatus associated with the Kramers operator described in Section 2.6 [refs. (6)–(9)]. Next, there is a great deal of extra background material which, although it is not exactly the same as that in the present book, preceded and aided the development of the irreducible tensor technique for finite groups. This material is very extensive, and I list only a few important early works which influenced me strongly [refs. (10)–(19)]. Fourth, there is the work of the Japanese school of ligand-field theorists [especially refs. (20)–(22)]. The present book owes a great deal to two of these papers [refs. (20), (21)] both explicitly, as in Chapters 7 and 10, and implicitly throughout. My own previous work on these group-theoretic techniques appears in refs. (23)–(26). Finally, I give various recent papers that contain interesting material which is related fairly closely to the mathematical techniques of the present book [refs. (27)–(36)]. Reference (36) defines coefficients, which are like W coefficients, but which involve also the two-valued representations. It includes a table of their values for the octahedral group. The references listed below are quoted by number earlier in the book.

REFERENCES

1. Condon, E. U. and G. H. Shortley, *The Theory of Atomic Spectra.* Cambridge University Press, 1953.

2. Edmonds, A. R., *Angular Momentum in Quantum Mechanics*. Princeton University Press, 1957.
3. Fano, U. and G. Racah, *Irreducible Tensorial Sets*. Academic Press, 1959.
4. Wigner, E. P., *Group Theory and Its Application to the Quantum Mechanics of Atomic Spectra*. Academic Press, 1959.
5. Jucys, A. P., I. B. Levinson, and V. V. Vanagas, *Mathematical Apparatus of the Theory of Angular Momenta* (in Russian). National publishers of political and scientific literature of the Lithuanian S. S. R., 1960.
6. Frobenius, G. and I. Schur, *Berliner Berichte*, 1906, p. 186.
7. Kramers, H. A., *K. Akad. van Wetenschappen*, **33** (1930), 959.
8. Wigner, E. P., *Nachr. Ges. Wiss. Göttingen*, (1932), p. 546.
9. Jahn, H. A., *Proc. Roy. Soc. A.*, **164** (1938), 117.
10. Bethe, H. A., *Ann. Phys. Leipzig*, **3** (1929), 133.
11. Jahn, H. A. and E. Teller, *Proc. Roy. Soc. A.*, **161** (1937), 220.
12. Murnaghan, F. D., *The Theory of Group Representations*. Baltimore: The Johns Hopkins Press, 1938.
13. Racah, G., *Phys. Rev.*, **61** (1942), 186.
14. Racah, G., *Phys. Rev.*, **62** (1942), 438.
15. Racah, G., *Phys. Rev.*, **63** (1943), 367.
16. Racah, G., *Phys. Rev.*, **76** (1949), 1352.
17. Araki, G., *Prog. Theor. Phys.*, **3** (1948), 152.
18. Abragam, A. and M. H. L. Pryce, *Proc. Roy. Soc. A.*, **205** (1951), 135.
19. Bleaney, B. and K. W. H. Stevens, *Rep. Prog. Phys.*, **16** (1953), 108.
20. Tanabe, Y. and S. Sugano, *J. Phys. Soc. Japan*, **9** (1954), 753.
21. Tanabe, Y. and H. Kamimura, *J. Phys. Soc. Japan*, **13** (1958), 394.
22. Tanabe, Y., *Prog. Theor. Phys.*, Supplement No. 14 (1960), 17.
23. Griffith, J. S., *Trans. Farad. Soc.*, **54** (1958), 1109.
24. Griffith, J. S., *Trans. Farad. Soc.*, **56** (1960), 193.
25. Griffith, J. S., *Molecular Physics*, **3** (1960), 79, 285, 457, 477.
26. Griffith, J. S., *The Theory of Transition-Metal Ions*. Cambridge University Press, 1961.
27. Koster, G. F., *Phys. Rev.*, **109** (1958), 227.
28. Jensen, L., *Phys. Rev.*, **110** (1958), 661.
29. Koster, G. F. and H. Statz, *Phys. Rev.*, **113** (1959), 445.
30. Jarrett, H. S., *J. Chem. Phys.*, **31** (1959), 1579.
31. Albrecht, A. C., *J. Chem. Phys.*, **33** (1960), 156, 169.
32. Parr, R. G., *J. Chem. Phys.*, **33** (1960), 1184.
33. Batarunas, I. V. and I. B. Levinson, *Trudy A. N. Nauk Litovskoj S. S. R.*, Serija B, **2(22)** (1960), 15.
34. McWeeny, R. and Y. Mizuno, *Proc. Roy. Soc. A.*, **259** (1961), 554.
35. Petruska, J., *J. Chem. Phys.*, **34** (1961), 1111.
36. Mauza, E. B. and I. V. Batarunas, *Trudy A. N. Nauk Litovskoj S. S. R.*, Serija B, **3(26)** (1961), 27.

Index

A CATALOG OF SELECTED

DOVER BOOKS

IN SCIENCE AND MATHEMATICS

Astronomy

BURNHAM'S CELESTIAL HANDBOOK, Robert Burnham, Jr. Thorough guide to the stars beyond our solar system. Exhaustive treatment. Alphabetical by constellation: Andromeda to Cetus in Vol. 1; Chamaeleon to Orion in Vol. 2; and Pavo to Vulpecula in Vol. 3. Hundreds of illustrations. Index in Vol. 3. 2,000pp. 6⅛ x 9¼.
Vol. I: 0-486-23567-X
Vol. II: 0-486-23568-8
Vol. III: 0-486-23673-0

EXPLORING THE MOON THROUGH BINOCULARS AND SMALL TELESCOPES, Ernest H. Cherrington, Jr. Informative, profusely illustrated guide to locating and identifying craters, rills, seas, mountains, other lunar features. Newly revised and updated with special section of new photos. Over 100 photos and diagrams. 240pp. 8¼ x 11. 0-486-24491-1

THE EXTRATERRESTRIAL LIFE DEBATE, 1750–1900, Michael J. Crowe. First detailed, scholarly study in English of the many ideas that developed from 1750 to 1900 regarding the existence of intelligent extraterrestrial life. Examines ideas of Kant, Herschel, Voltaire, Percival Lowell, many other scientists and thinkers. 16 illustrations. 704pp. 5⅜ x 8½. 0-486-40675-X

THEORIES OF THE WORLD FROM ANTIQUITY TO THE COPERNICAN REVOLUTION, Michael J. Crowe. Newly revised edition of an accessible, enlightening book recreates the change from an earth-centered to a sun-centered conception of the solar system. 242pp. 5⅜ x 8½. 0-486-41444-2

A HISTORY OF ASTRONOMY, A. Pannekoek. Well-balanced, carefully reasoned study covers such topics as Ptolemaic theory, work of Copernicus, Kepler, Newton, Eddington's work on stars, much more. Illustrated. References. 521pp. 5⅜ x 8½.
0-486-65994-1

A COMPLETE MANUAL OF AMATEUR ASTRONOMY: TOOLS AND TECHNIQUES FOR ASTRONOMICAL OBSERVATIONS, P. Clay Sherrod with Thomas L. Koed. Concise, highly readable book discusses: selecting, setting up and maintaining a telescope; amateur studies of the sun; lunar topography and occultations; observations of Mars, Jupiter, Saturn, the minor planets and the stars; an introduction to photoelectric photometry; more. 1981 ed. 124 figures. 25 halftones. 37 tables. 335pp. 6½ x 9¼. 0-486-40675-X

AMATEUR ASTRONOMER'S HANDBOOK, J. B. Sidgwick. Timeless, comprehensive coverage of telescopes, mirrors, lenses, mountings, telescope drives, micrometers, spectroscopes, more. 189 illustrations. 576pp. 5⅝ x 8¼. (Available in U.S. only.)
0-486-24034-7

STARS AND RELATIVITY, Ya. B. Zel'dovich and I. D. Novikov. Vol. 1 of *Relativistic Astrophysics* by famed Russian scientists. General relativity, properties of matter under astrophysical conditions, stars, and stellar systems. Deep physical insights, clear presentation. 1971 edition. References. 544pp. 5⅝ x 8¼. 0-486-69424-0

Chemistry

THE SCEPTICAL CHYMIST: THE CLASSIC 1661 TEXT, Robert Boyle. Boyle defines the term "element," asserting that all natural phenomena can be explained by the motion and organization of primary particles. 1911 ed. viii+232pp. 5⅜ x 8½. 0-486-42825-7

RADIOACTIVE SUBSTANCES, Marie Curie. Here is the celebrated scientist's doctoral thesis, the prelude to her receipt of the 1903 Nobel Prize. Curie discusses establishing atomic character of radioactivity found in compounds of uranium and thorium; extraction from pitchblende of polonium and radium; isolation of pure radium chloride; determination of atomic weight of radium; plus electric, photographic, luminous, heat, color effects of radioactivity. ii+94pp. 5⅜ x 8½. 0-486-42550-9

CHEMICAL MAGIC, Leonard A. Ford. Second Edition, Revised by E. Winston Grundmeier. Over 100 unusual stunts demonstrating cold fire, dust explosions, much more. Text explains scientific principles and stresses safety precautions. 128pp. 5⅜ x 8½. 0-486-67628-5

THE DEVELOPMENT OF MODERN CHEMISTRY, Aaron J. Ihde. Authoritative history of chemistry from ancient Greek theory to 20th-century innovation. Covers major chemists and their discoveries. 209 illustrations. 14 tables. Bibliographies. Indices. Appendices. 851pp. 5⅜ x 8½. 0-486-64235-6

CATALYSIS IN CHEMISTRY AND ENZYMOLOGY, William P. Jencks. Exceptionally clear coverage of mechanisms for catalysis, forces in aqueous solution, carbonyl- and acyl-group reactions, practical kinetics, more. 864pp. 5⅜ x 8½. 0-486-65460-5

ELEMENTS OF CHEMISTRY, Antoine Lavoisier. Monumental classic by founder of modern chemistry in remarkable reprint of rare 1790 Kerr translation. A must for every student of chemistry or the history of science. 539pp. 5⅜ x 8½. 0-486-64624-6

THE HISTORICAL BACKGROUND OF CHEMISTRY, Henry M. Leicester. Evolution of ideas, not individual biography. Concentrates on formulation of a coherent set of chemical laws. 260pp. 5⅜ x 8½. 0-486-61053-5

A SHORT HISTORY OF CHEMISTRY, J. R. Partington. Classic exposition explores origins of chemistry, alchemy, early medical chemistry, nature of atmosphere, theory of valency, laws and structure of atomic theory, much more. 428pp. 5⅜ x 8½. (Available in U.S. only.) 0-486-65977-1

GENERAL CHEMISTRY, Linus Pauling. Revised 3rd edition of classic first-year text by Nobel laureate. Atomic and molecular structure, quantum mechanics, statistical mechanics, thermodynamics correlated with descriptive chemistry. Problems. 992pp. 5⅜ x 8½. 0-486-65622-5

FROM ALCHEMY TO CHEMISTRY, John Read. Broad, humanistic treatment focuses on great figures of chemistry and ideas that revolutionized the science. 50 illustrations. 240pp. 5⅜ x 8½. 0-486-28690-8

Engineering

DE RE METALLICA, Georgius Agricola. The famous Hoover translation of greatest treatise on technological chemistry, engineering, geology, mining of early modern times (1556). All 289 original woodcuts. 638pp. 6¾ x 11. 0-486-60006-8

FUNDAMENTALS OF ASTRODYNAMICS, Roger Bate et al. Modern approach developed by U.S. Air Force Academy. Designed as a first course. Problems, exercises. Numerous illustrations. 455pp. 5⅜ x 8½. 0-486-60061-0

DYNAMICS OF FLUIDS IN POROUS MEDIA, Jacob Bear. For advanced students of ground water hydrology, soil mechanics and physics, drainage and irrigation engineering and more. 335 illustrations. Exercises, with answers. 784pp. 6⅛ x 9¼. 0-486-65675-6

THEORY OF VISCOELASTICITY (Second Edition), Richard M. Christensen. Complete consistent description of the linear theory of the viscoelastic behavior of materials. Problem-solving techniques discussed. 1982 edition. 29 figures. xiv+364pp. 6⅛ x 9¼. 0-486-42880-X

MECHANICS, J. P. Den Hartog. A classic introductory text or refresher. Hundreds of applications and design problems illuminate fundamentals of trusses, loaded beams and cables, etc. 334 answered problems. 462pp. 5⅜ x 8½. 0-486-60754-2

MECHANICAL VIBRATIONS, J. P. Den Hartog. Classic textbook offers lucid explanations and illustrative models, applying theories of vibrations to a variety of practical industrial engineering problems. Numerous figures. 233 problems, solutions. Appendix. Index. Preface. 436pp. 5⅜ x 8½. 0-486-64785-4

STRENGTH OF MATERIALS, J. P. Den Hartog. Full, clear treatment of basic material (tension, torsion, bending, etc.) plus advanced material on engineering methods, applications. 350 answered problems. 323pp. 5⅜ x 8½. 0-486-60755-0

A HISTORY OF MECHANICS, René Dugas. Monumental study of mechanical principles from antiquity to quantum mechanics. Contributions of ancient Greeks, Galileo, Leonardo, Kepler, Lagrange, many others. 671pp. 5⅜ x 8½. 0-486-65632-2

STABILITY THEORY AND ITS APPLICATIONS TO STRUCTURAL MECHANICS, Clive L. Dym. Self-contained text focuses on Koiter postbuckling analyses, with mathematical notions of stability of motion. Basing minimum energy principles for static stability upon dynamic concepts of stability of motion, it develops asymptotic buckling and postbuckling analyses from potential energy considerations, with applications to columns, plates, and arches. 1974 ed. 208pp. 5⅜ x 8½. 0-486-42541-X

METAL FATIGUE, N. E. Frost, K. J. Marsh, and L. P. Pook. Definitive, clearly written, and well-illustrated volume addresses all aspects of the subject, from the historical development of understanding metal fatigue to vital concepts of the cyclic stress that causes a crack to grow. Includes 7 appendixes. 544pp. 5⅜ x 8½. 0-486-40927-9

ROCKETS, Robert Goddard. Two of the most significant publications in the history of rocketry and jet propulsion: "A Method of Reaching Extreme Altitudes" (1919) and "Liquid Propellant Rocket Development" (1936). 128pp. 5⅜ x 8½. 0-486-42537-1

STATISTICAL MECHANICS: PRINCIPLES AND APPLICATIONS, Terrell L. Hill. Standard text covers fundamentals of statistical mechanics, applications to fluctuation theory, imperfect gases, distribution functions, more. 448pp. 5⅜ x 8½. 0-486-65390-0

ENGINEERING AND TECHNOLOGY 1650–1750: ILLUSTRATIONS AND TEXTS FROM ORIGINAL SOURCES, Martin Jensen. Highly readable text with more than 200 contemporary drawings and detailed engravings of engineering projects dealing with surveying, leveling, materials, hand tools, lifting equipment, transport and erection, piling, bailing, water supply, hydraulic engineering, and more. Among the specific projects outlined-transporting a 50-ton stone to the Louvre, erecting an obelisk, building timber locks, and dredging canals. 207pp. 8⅜ x 11¼. 0-486-42232-1

THE VARIATIONAL PRINCIPLES OF MECHANICS, Cornelius Lanczos. Graduate level coverage of calculus of variations, equations of motion, relativistic mechanics, more. First inexpensive paperbound edition of classic treatise. Index. Bibliography. 418pp. 5⅜ x 8½. 0-486-65067-7

PROTECTION OF ELECTRONIC CIRCUITS FROM OVERVOLTAGES, Ronald B. Standler. Five-part treatment presents practical rules and strategies for circuits designed to protect electronic systems from damage by transient overvoltages. 1989 ed. xxiv+434pp. 6⅛ x 9¼. 0-486-42552-5

ROTARY WING AERODYNAMICS, W. Z. Stepniewski. Clear, concise text covers aerodynamic phenomena of the rotor and offers guidelines for helicopter performance evaluation. Originally prepared for NASA. 537 figures. 640pp. 6⅛ x 9¼. 0-486-64647-5

INTRODUCTION TO SPACE DYNAMICS, William Tyrrell Thomson. Comprehensive, classic introduction to space-flight engineering for advanced undergraduate and graduate students. Includes vector algebra, kinematics, transformation of coordinates. Bibliography. Index. 352pp. 5⅜ x 8½. 0-486-65113-4

HISTORY OF STRENGTH OF MATERIALS, Stephen P. Timoshenko. Excellent historical survey of the strength of materials with many references to the theories of elasticity and structure. 245 figures. 452pp. 5⅜ x 8½. 0-486-61187-6

ANALYTICAL FRACTURE MECHANICS, David J. Unger. Self-contained text supplements standard fracture mechanics texts by focusing on analytical methods for determining crack-tip stress and strain fields. 336pp. 6⅛ x 9¼. 0-486-41737-9

STATISTICAL MECHANICS OF ELASTICITY, J. H. Weiner. Advanced, self-contained treatment illustrates general principles and elastic behavior of solids. Part 1, based on classical mechanics, studies thermoelastic behavior of crystalline and polymeric solids. Part 2, based on quantum mechanics, focuses on interatomic force laws, behavior of solids, and thermally activated processes. For students of physics and chemistry and for polymer physicists. 1983 ed. 96 figures. 496pp. 5⅜ x 8½. 0-486-42260-7

Mathematics

FUNCTIONAL ANALYSIS (Second Corrected Edition), George Bachman and Lawrence Narici. Excellent treatment of subject geared toward students with background in linear algebra, advanced calculus, physics and engineering. Text covers introduction to inner-product spaces, normed, metric spaces, and topological spaces; complete orthonormal sets, the Hahn-Banach Theorem and its consequences, and many other related subjects. 1966 ed. 544pp. 6⅛ x 9¼. 0-486-40251-7

ASYMPTOTIC EXPANSIONS OF INTEGRALS, Norman Bleistein & Richard A. Handelsman. Best introduction to important field with applications in a variety of scientific disciplines. New preface. Problems. Diagrams. Tables. Bibliography. Index. 448pp. 5⅜ x 8½. 0-486-65082-0

VECTOR AND TENSOR ANALYSIS WITH APPLICATIONS, A. I. Borisenko and I. E. Tarapov. Concise introduction. Worked-out problems, solutions, exercises. 257pp. 5⅝ x 8¼. 0-486-63833-2

AN INTRODUCTION TO ORDINARY DIFFERENTIAL EQUATIONS, Earl A. Coddington. A thorough and systematic first course in elementary differential equations for undergraduates in mathematics and science, with many exercises and problems (with answers). Index. 304pp. 5⅜ x 8½. 0-486-65942-9

FOURIER SERIES AND ORTHOGONAL FUNCTIONS, Harry F. Davis. An incisive text combining theory and practical example to introduce Fourier series, orthogonal functions and applications of the Fourier method to boundary-value problems. 570 exercises. Answers and notes. 416pp. 5⅜ x 8½. 0-486-65973-9

COMPUTABILITY AND UNSOLVABILITY, Martin Davis. Classic graduate-level introduction to theory of computability, usually referred to as theory of recurrent functions. New preface and appendix. 288pp. 5⅜ x 8½. 0-486-61471-9

ASYMPTOTIC METHODS IN ANALYSIS, N. G. de Bruijn. An inexpensive, comprehensive guide to asymptotic methods–the pioneering work that teaches by explaining worked examples in detail. Index. 224pp. 5⅜ x 8½ 0-486-64221-6

APPLIED COMPLEX VARIABLES, John W. Dettman. Step-by-step coverage of fundamentals of analytic function theory–plus lucid exposition of five important applications: Potential Theory; Ordinary Differential Equations; Fourier Transforms; Laplace Transforms; Asymptotic Expansions. 66 figures. Exercises at chapter ends. 512pp. 5⅜ x 8½. 0-486-64670-X

INTRODUCTION TO LINEAR ALGEBRA AND DIFFERENTIAL EQUATIONS, John W. Dettman. Excellent text covers complex numbers, determinants, orthonormal bases, Laplace transforms, much more. Exercises with solutions. Undergraduate level. 416pp. 5⅜ x 8½. 0-486-65191-6

RIEMANN'S ZETA FUNCTION, H. M. Edwards. Superb, high-level study of landmark 1859 publication entitled "On the Number of Primes Less Than a Given Magnitude" traces developments in mathematical theory that it inspired. xiv+315pp. 5⅜ x 8½. 0-486-41740-9

TENSOR CALCULUS, J.L. Synge and A. Schild. Widely used introductory text covers spaces and tensors, basic operations in Riemannian space, non-Riemannian spaces, etc. 324pp. 5⅜ x 8¼. 0-486-63612-7

ORDINARY DIFFERENTIAL EQUATIONS, Morris Tenenbaum and Harry Pollard. Exhaustive survey of ordinary differential equations for undergraduates in mathematics, engineering, science. Thorough analysis of theorems. Diagrams. Bibliography. Index. 818pp. 5⅜ x 8½. 0-486-64940-7

INTEGRAL EQUATIONS, F. G. Tricomi. Authoritative, well-written treatment of extremely useful mathematical tool with wide applications. Volterra Equations, Fredholm Equations, much more. Advanced undergraduate to graduate level. Exercises. Bibliography. 238pp. 5⅜ x 8½. 0-486-64828-1

FOURIER SERIES, Georgi P. Tolstov. Translated by Richard A. Silverman. A valuable addition to the literature on the subject, moving clearly from subject to subject and theorem to theorem. 107 problems, answers. 336pp. 5⅜ x 8½. 0-486-63317-9

INTRODUCTION TO MATHEMATICAL THINKING, Friedrich Waismann. Examinations of arithmetic, geometry, and theory of integers; rational and natural numbers; complete induction; limit and point of accumulation; remarkable curves; complex and hypercomplex numbers, more. 1959 ed. 27 figures. xii+260pp. 5⅜ x 8½. 0-486-63317-9

POPULAR LECTURES ON MATHEMATICAL LOGIC, Hao Wang. Noted logician's lucid treatment of historical developments, set theory, model theory, recursion theory and constructivism, proof theory, more. 3 appendixes. Bibliography. 1981 edition. ix + 283pp. 5⅜ x 8½. 0-486-67632-3

CALCULUS OF VARIATIONS, Robert Weinstock. Basic introduction covering isoperimetric problems, theory of elasticity, quantum mechanics, electrostatics, etc. Exercises throughout. 326pp. 5⅜ x 8½. 0-486-63069-2

THE CONTINUUM: A CRITICAL EXAMINATION OF THE FOUNDATION OF ANALYSIS, Hermann Weyl. Classic of 20th-century foundational research deals with the conceptual problem posed by the continuum. 156pp. 5⅜ x 8½. 0-486-67982-9

CHALLENGING MATHEMATICAL PROBLEMS WITH ELEMENTARY SOLUTIONS, A. M. Yaglom and I. M. Yaglom. Over 170 challenging problems on probability theory, combinatorial analysis, points and lines, topology, convex polygons, many other topics. Solutions. Total of 445pp. 5⅜ x 8½. Two-vol. set. Vol. I: 0-486-65536-9 Vol. II: 0-486-65537-7

INTRODUCTION TO PARTIAL DIFFERENTIAL EQUATIONS WITH APPLICATIONS, E. C. Zachmanoglou and Dale W. Thoe. Essentials of partial differential equations applied to common problems in engineering and the physical sciences. Problems and answers. 416pp. 5⅜ x 8½. 0-486-65251-3

THE THEORY OF GROUPS, Hans J. Zassenhaus. Well-written graduate-level text acquaints reader with group-theoretic methods and demonstrates their usefulness in mathematics. Axioms, the calculus of complexes, homomorphic mapping, *p*-group theory, more. 276pp. 5⅜ x 8½. 0-486-40922-8

Math–Decision Theory, Statistics, Probability

ELEMENTARY DECISION THEORY, Herman Chernoff and Lincoln E. Moses. Clear introduction to statistics and statistical theory covers data processing, probability and random variables, testing hypotheses, much more. Exercises. 364pp. 5⅜ x 8½. 0-486-65218-1

STATISTICS MANUAL, Edwin L. Crow et al. Comprehensive, practical collection of classical and modern methods prepared by U.S. Naval Ordnance Test Station. Stress on use. Basics of statistics assumed. 288pp. 5⅜ x 8½. 0-486-60599-X

SOME THEORY OF SAMPLING, William Edwards Deming. Analysis of the problems, theory and design of sampling techniques for social scientists, industrial managers and others who find statistics important at work. 61 tables. 90 figures. xvii +602pp. 5⅜ x 8½. 0-486-64684-X

LINEAR PROGRAMMING AND ECONOMIC ANALYSIS, Robert Dorfman, Paul A. Samuelson and Robert M. Solow. First comprehensive treatment of linear programming in standard economic analysis. Game theory, modern welfare economics, Leontief input-output, more. 525pp. 5⅜ x 8½. 0-486-65491-5

PROBABILITY: AN INTRODUCTION, Samuel Goldberg. Excellent basic text covers set theory, probability theory for finite sample spaces, binomial theorem, much more. 360 problems. Bibliographies. 322pp. 5⅜ x 8½. 0-486-65252-1

GAMES AND DECISIONS: INTRODUCTION AND CRITICAL SURVEY, R. Duncan Luce and Howard Raiffa. Superb nontechnical introduction to game theory, primarily applied to social sciences. Utility theory, zero-sum games, n-person games, decision-making, much more. Bibliography. 509pp. 5⅜ x 8½. 0-486-65943-7

INTRODUCTION TO THE THEORY OF GAMES, J. C. C. McKinsey. This comprehensive overview of the mathematical theory of games illustrates applications to situations involving conflicts of interest, including economic, social, political, and military contexts. Appropriate for advanced undergraduate and graduate courses; advanced calculus a prerequisite. 1952 ed. x+372pp. 5⅜ x 8½. 0-486-42811-7

FIFTY CHALLENGING PROBLEMS IN PROBABILITY WITH SOLUTIONS, Frederick Mosteller. Remarkable puzzlers, graded in difficulty, illustrate elementary and advanced aspects of probability. Detailed solutions. 88pp. 5⅜ x 8½. 65355-2

PROBABILITY THEORY: A CONCISE COURSE, Y. A. Rozanov. Highly readable, self-contained introduction covers combination of events, dependent events, Bernoulli trials, etc. 148pp. 5⅜ x 8¼. 0-486-63544-9

STATISTICAL METHOD FROM THE VIEWPOINT OF QUALITY CONTROL, Walter A. Shewhart. Important text explains regulation of variables, uses of statistical control to achieve quality control in industry, agriculture, other areas. 192pp. 5⅜ x 8½. 0-486-65232-7

Math–Geometry and Topology

ELEMENTARY CONCEPTS OF TOPOLOGY, Paul Alexandroff. Elegant, intuitive approach to topology from set-theoretic topology to Betti groups; how concepts of topology are useful in math and physics. 25 figures. 57pp. $5\frac{3}{8}$ x $8\frac{1}{2}$. 0-486-60747-X

COMBINATORIAL TOPOLOGY, P. S. Alexandrov. Clearly written, well-organized, three-part text begins by dealing with certain classic problems without using the formal techniques of homology theory and advances to the central concept, the Betti groups. Numerous detailed examples. 654pp. $5\frac{3}{8}$ x $8\frac{1}{2}$. 0-486-40179-0

EXPERIMENTS IN TOPOLOGY, Stephen Barr. Classic, lively explanation of one of the byways of mathematics. Klein bottles, Moebius strips, projective planes, map coloring, problem of the Koenigsberg bridges, much more, described with clarity and wit. 43 figures. 210pp. $5\frac{3}{8}$ x $8\frac{1}{2}$. 0-486-25933-1

THE GEOMETRY OF RENÉ DESCARTES, René Descartes. The great work founded analytical geometry. Original French text, Descartes's own diagrams, together with definitive Smith-Latham translation. 244pp. $5\frac{3}{8}$ x $8\frac{1}{2}$. 0-486-60068-8

EUCLIDEAN GEOMETRY AND TRANSFORMATIONS, Clayton W. Dodge. This introduction to Euclidean geometry emphasizes transformations, particularly isometries and similarities. Suitable for undergraduate courses, it includes numerous examples, many with detailed answers. 1972 ed. viii+296pp. $6\frac{1}{8}$ x $9\frac{1}{4}$. 0-486-43476-1

PRACTICAL CONIC SECTIONS: THE GEOMETRIC PROPERTIES OF ELLIPSES, PARABOLAS AND HYPERBOLAS, J. W. Downs. This text shows how to create ellipses, parabolas, and hyperbolas. It also presents historical background on their ancient origins and describes the reflective properties and roles of curves in design applications. 1993 ed. 98 figures. xii+100pp. $6\frac{1}{2}$ x $9\frac{1}{4}$. 0-486-42876-1

THE THIRTEEN BOOKS OF EUCLID'S ELEMENTS, translated with introduction and commentary by Sir Thomas L. Heath. Definitive edition. Textual and linguistic notes, mathematical analysis. 2,500 years of critical commentary. Unabridged. 1,414pp. $5\frac{3}{8}$ x $8\frac{1}{2}$. Three-vol. set.
Vol. I: 0-486-60088-2 Vol. II: 0-486-60089-0 Vol. III: 0-486-60090-4

SPACE AND GEOMETRY: IN THE LIGHT OF PHYSIOLOGICAL, PSYCHOLOGICAL AND PHYSICAL INQUIRY, Ernst Mach. Three essays by an eminent philosopher and scientist explore the nature, origin, and development of our concepts of space, with a distinctness and precision suitable for undergraduate students and other readers. 1906 ed. vi+148pp. $5\frac{3}{8}$ x $8\frac{1}{2}$. 0-486-43909-7

GEOMETRY OF COMPLEX NUMBERS, Hans Schwerdtfeger. Illuminating, widely praised book on analytic geometry of circles, the Moebius transformation, and two-dimensional non-Euclidean geometries. 200pp. $5\frac{5}{8}$ x $8\frac{1}{4}$. 0-486-63830-8

DIFFERENTIAL GEOMETRY, Heinrich W. Guggenheimer. Local differential geometry as an application of advanced calculus and linear algebra. Curvature, transformation groups, surfaces, more. Exercises. 62 figures. 378pp. $5\frac{3}{8}$ x $8\frac{1}{2}$. 0-486-63433-7

History of Math

THE WORKS OF ARCHIMEDES, Archimedes (T. L. Heath, ed.). Topics include the famous problems of the ratio of the areas of a cylinder and an inscribed sphere; the measurement of a circle; the properties of conoids, spheroids, and spirals; and the quadrature of the parabola. Informative introduction. clxxxvi+326pp. 5⅜ x 8½. 0-486-42084-1

A SHORT ACCOUNT OF THE HISTORY OF MATHEMATICS, W. W. Rouse Ball. One of clearest, most authoritative surveys from the Egyptians and Phoenicians through 19th-century figures such as Grassman, Galois, Riemann. Fourth edition. 522pp. 5⅜ x 8½. 0-486-20630-0

THE HISTORY OF THE CALCULUS AND ITS CONCEPTUAL DEVELOPMENT, Carl B. Boyer. Origins in antiquity, medieval contributions, work of Newton, Leibniz, rigorous formulation. Treatment is verbal. 346pp. 5⅜ x 8½. 0-486-60509-4

THE HISTORICAL ROOTS OF ELEMENTARY MATHEMATICS, Lucas N. H. Bunt, Phillip S. Jones, and Jack D. Bedient. Fundamental underpinnings of modern arithmetic, algebra, geometry and number systems derived from ancient civilizations. 320pp. 5⅜ x 8½. 0-486-25563-8

A HISTORY OF MATHEMATICAL NOTATIONS, Florian Cajori. This classic study notes the first appearance of a mathematical symbol and its origin, the competition it encountered, its spread among writers in different countries, its rise to popularity, its eventual decline or ultimate survival. Original 1929 two-volume edition presented here in one volume. xxviii+820pp. 5⅜ x 8½. 0-486-67766-4

GAMES, GODS & GAMBLING: A HISTORY OF PROBABILITY AND STATISTICAL IDEAS, F. N. David. Episodes from the lives of Galileo, Fermat, Pascal, and others illustrate this fascinating account of the roots of mathematics. Features thought-provoking references to classics, archaeology, biography, poetry. 1962 edition. 304pp. 5⅜ x 8½. (Available in U.S. only.) 0-486-40023-9

OF MEN AND NUMBERS: THE STORY OF THE GREAT MATHEMATICIANS, Jane Muir. Fascinating accounts of the lives and accomplishments of history's greatest mathematical minds–Pythagoras, Descartes, Euler, Pascal, Cantor, many more. Anecdotal, illuminating. 30 diagrams. Bibliography. 256pp. 5⅜ x 8½. 0-486-28973-7

HISTORY OF MATHEMATICS, David E. Smith. Nontechnical survey from ancient Greece and Orient to late 19th century; evolution of arithmetic, geometry, trigonometry, calculating devices, algebra, the calculus. 362 illustrations. 1,355pp. 5⅜ x 8½. Two-vol. set. Vol. I: 0-486-20429-4 Vol. II: 0-486-20430-8

A CONCISE HISTORY OF MATHEMATICS, Dirk J. Struik. The best brief history of mathematics. Stresses origins and covers every major figure from ancient Near East to 19th century. 41 illustrations. 195pp. 5⅜ x 8½. 0-486-60255-9

Physics

OPTICAL RESONANCE AND TWO-LEVEL ATOMS, L. Allen and J. H. Eberly. Clear, comprehensive introduction to basic principles behind all quantum optical resonance phenomena. 53 illustrations. Preface. Index. 256pp. 5⅜ x 8½. 0-486-65533-4

QUANTUM THEORY, David Bohm. This advanced undergraduate-level text presents the quantum theory in terms of qualitative and imaginative concepts, followed by specific applications worked out in mathematical detail. Preface. Index. 655pp. 5⅜ x 8½. 0-486-65969-0

ATOMIC PHYSICS (8th EDITION), Max Born. Nobel laureate's lucid treatment of kinetic theory of gases, elementary particles, nuclear atom, wave-corpuscles, atomic structure and spectral lines, much more. Over 40 appendices, bibliography. 495pp. 5⅜ x 8½. 0-486-65984-4

A SOPHISTICATE'S PRIMER OF RELATIVITY, P. W. Bridgman. Geared toward readers already acquainted with special relativity, this book transcends the view of theory as a working tool to answer natural questions: What is a frame of reference? What is a "law of nature"? What is the role of the "observer"? Extensive treatment, written in terms accessible to those without a scientific background. 1983 ed. xlviii+172pp. 5⅜ x 8½. 0-486-42549-5

AN INTRODUCTION TO HAMILTONIAN OPTICS, H. A. Buchdahl. Detailed account of the Hamiltonian treatment of aberration theory in geometrical optics. Many classes of optical systems defined in terms of the symmetries they possess. Problems with detailed solutions. 1970 edition. xv + 360pp. 5⅜ x 8½. 0-486-67597-1

PRIMER OF QUANTUM MECHANICS, Marvin Chester. Introductory text examines the classical quantum bead on a track: its state and representations; operator eigenvalues; harmonic oscillator and bound bead in a symmetric force field; and bead in a spherical shell. Other topics include spin, matrices, and the structure of quantum mechanics; the simplest atom; indistinguishable particles; and stationary-state perturbation theory. 1992 ed. xiv+314pp. 6⅛ x 9¼. 0-486-42878-8

LECTURES ON QUANTUM MECHANICS, Paul A. M. Dirac. Four concise, brilliant lectures on mathematical methods in quantum mechanics from Nobel Prize-winning quantum pioneer build on idea of visualizing quantum theory through the use of classical mechanics. 96pp. 5⅜ x 8½. 0-486-41713-1

THIRTY YEARS THAT SHOOK PHYSICS: THE STORY OF QUANTUM THEORY, George Gamow. Lucid, accessible introduction to influential theory of energy and matter. Careful explanations of Dirac's anti-particles, Bohr's model of the atom, much more. 12 plates. Numerous drawings. 240pp. 5⅜ x 8½. 0-486-24895-X

ELECTRONIC STRUCTURE AND THE PROPERTIES OF SOLIDS: THE PHYSICS OF THE CHEMICAL BOND, Walter A. Harrison. Innovative text offers basic understanding of the electronic structure of covalent and ionic solids, simple metals, transition metals and their compounds. Problems. 1980 edition. 582pp. 6⅛ x 9¼. 0-486-66021-4

CATALOG OF DOVER BOOKS